P9-DHJ-968

The Bird Watching Answer Book

The **Cornell** Lab of Ornithology

THE Bird Watching

ANSWER BOOK

Everything You Need to Know to
Enjoy Birds in Your Backyard and Beyond

LAURA ERICKSON

Storey Publishing

*The mission of Storey Publishing is to serve our customers by
publishing practical information that encourages
personal independence in harmony with the environment.*

Edited by Deborah Balmuth and Lisa H. Hiley
Art direction and book design by Mary Winkelman Velgos
Text production by Jennifer Jepson Smith

Illustrations by © Pedro Fernandes

Indexed by Nancy D. Wood

    The information in this book is true and complete to the best of our knowledge. All recom-
mendations are made without guarantee on the part of the author or Storey Publishing. The
author and publisher disclaim any liability in connection with the use of this information.
    Storey books are available for special premium and promotional uses and for customized
editions. For further information, please call 1-800-793-9396.

Storey Publishing
210 MASS MoCA Way
North Adams, MA 01247
www.storey.com

Printed in China by Regent Publishing Services
10 9 8 7 6 5 4 3 2

LIBRARY OF CONGRESS CATALOGING-IN-PUBLICATION DATA
Erickson, Laura, 1951–
   The bird watching answer book / Laura Erickson.
      p. cm.
   Includes index.
   ISBN 978-1-60342-452-3 (flexibind with paper spine : alk. paper)
   1. Bird watching. I. Title. QL677.5.E75 2009
598.072'34—dc22

                                              2009023708

*For my children, and for people everywhere
whose minds bubble over with questions.*

Acknowledgments

This book would not have been possible without the resources and people at the Cornell Lab of Ornithology. I've been ever amazed at how accessible and willing to answer my questions the scientists here are. I'd especially like to thank the many experts who looked over the manuscript and made suggestions; their help improved the book enormously. Any errors that remain are entirely my fault.

I'd especially like to acknowledge Miyoko Chu, the director of communications at the Lab. Her eagerness to share the Lab's many scientific and educational resources with people everywhere is exceeded only by her commitment to excellence. She has been one of the most inspiring, dedicated, and helpful mentors I've ever had.

The Cornell Lab of Ornithology is a nonprofit, member-supported organization that solves critical problems facing birds and other wildlife by using the best science and technology — and by inspiring people of all ages and backgrounds to care about and protect the planet. For more information, visit *www.birds.cornell.edu.*

Contents

PART TWO
Bird Brains: Avian Behavior and Intelligence

PART THREE
All About Birds, Inside and Out

Introduction

"Why is that Blue Jay bald?" "Why is a woodpecker digging holes in my house?" "What do I need to do to attract cardinals to my yard?" "Where did all these geese come from?" "How could bird feathers convict a murderer?"

I've been writing about birds for the past 30 years and have been speaking about them on the radio for more than 20 years, and from the very beginning, people have been plying me with questions. Birds are beautiful and fascinating, some easily seen in our own backyards, others worth making long trips to exotic places just to steal a glimpse. Based on a U.S. Fish and Wildlife Service survey conducted in 2005, more than 71 million Americans watch wildlife, an 8 percent increase since 2000. During 2005 alone, these wildlife watchers spent more than $45 billion on activities described as "closely observing, photographing, and/or feeding wildlife." Millions of people clicked on a single YouTube video showing a Neotropical bird, a Red-capped Manakin, performing its "moonwalk" courtship display. As the world grows ever more computerized and mechanized, we grow hungrier to experience nature, and perhaps especially hungry to experience birdlife.

Of all wildlife, birds are the animals we notice most in our daily lives. Even if we don't have feeders and don't pay much attention to nature, we can't help but notice a cardinal's song ringing through the air on a frosty February morning, ducks floating on the pond in the park, geese blocking passage on the golf course green. Over 700 species of wild birds breed in North

America, from hummingbirds to eagles, and prairie-chickens to city pigeons. Each species is unique, and when we meet up with an unfamiliar one, or with a familiar bird doing an unfamiliar behavior, we can't help but wonder.

Some of our encounters with birds are especially magical. Discovering a robin nest on our windowsill. Coming upon a tiny hummingbird doing his courtship flight, making deep swoops and dives, his wings buzzing mightily. Gazing at tens of thousands of cranes descending against a sunset sky to sleep on the Platte River in Nebraska. Watching from a boat as dozens of Atlantic Puffins fly overhead to their nesting sites, each one holding as many as 12 tiny fish lined up in its beak. Cooking breakfast on a picnic table at Yellowstone National Park and having Gray Jays and Steller's Jays alight on the hot stove to share some bacon.

Whether our experiences with birds are homey or exotic, beautiful or bizarre, thrilling or even a bit unpleasant, they fill us with wonder, and with questions. With the help of scientists and resources at the Cornell Lab of Ornithology, I tackle some of these bird questions anew. I hope this book answers most of your questions and inspires you to spend more time experiencing birds in nature, to come up with more questions, and to discover new answers yourself!

— LAURA ERICKSON

PART ONE

For the Birds: Feeding, Watching, and Protecting Our Feathered Friends

CHAPTER 1

Guess Who's Coming to Dinner? The Art of Bird Feeding

In 2006, Americans spent about $4 billion on bird seed and other food for wild birds and other animals. To put this in perspective, that same year they spent $7.1 billion on plasma TV sets, $11 billion on bottled water, $13.5 billion on home video and computer games, $24.1 billion on DVD rentals and purchases, and $58.5 billion on weddings. But that's still a lot of bird seed!

People who spend their hard-earned dollars feeding birds want to ensure that they are helping, not hurting, them. And naturally we also want to provide food for the birds we most enjoy watching, like chickadees and cardinals. Many people consider squirrels and black-birds to be nuisances. Some communities fine people if pigeons visit their feeders. Small wonder that I get hundreds of questions about bird feeding every year!

^v^v^v^v^v^v^v^v^v^v^v^v^v^v^v^v^v^v^v^v^v^v^v^v^v^v^v^v

Feeding Backyard Birds

Q I love watching the birds that visit my feeders in winter, but I worry that the birds might become too dependent on me. Is it okay to keep feeding them?

A Yes. A study of 348 color-banded Black-capped Chicka-dees in Wisconsin found that the birds took only about 21 percent of their daily energy from feeders. The study concluded that although natural food supplied most of their needs, feeders provided an important supplement to their natural diet.

The same researchers found that Black-capped Chickadees with access to feeders were more likely to survive very harsh winters. So your feeders can make tough times easier for the birds, but even when they visit feeders regularly, the birds still know how to find natural sources of food.

∨ ∨

Q **It seems like a bad idea to feed birds during spring, summer, and fall. Doesn't it turn them into moochers and mess up their urge to migrate?**

A Nope! The main cue that urges birds to migrate is the change in day length during spring and fall. Studies show that this instinct is so strong that when migratory birds are temporarily held in captivity with plenty of food, they become restless when it's time to migrate. They even fly against the side of the cage facing the direction they must travel.

Migrating birds are typically in a hurry to reach their breeding grounds in spring or to travel to their wintering grounds in fall. Even if they find your feeder along the way, most birds typically won't stop for more than a few days unless weather conditions are dire or they've lost a critical amount of weight and need to fatten up. During especially bad weather, feeders may mean the difference between life and death for some of these birds.

During spells of bad weather over the course of spring migration, when there may be little food available before the burst of plant growth and emerging insects, birds that seldom come to feeders, such as warblers and tanagers, may visit, particularly for suet. Keeping feeders active during spring and fall will attract local birds that are searching for the most reliable feeding areas, and their presence will entice migrants passing through your neighborhood to visit.

At times, a lost or injured bird may show up at your feeder, but remember that the injury or mix-up in the bird's migratory instinct wasn't *caused* by your feeders. Your feeders can buy some time for these individuals and give them a better chance to move on when they are ready.

Some people stop feeding birds throughout the summer months when other sources of food are more abundant. However, the nesting season demands a lot of energy from birds

FEEDING AN ADDICTION (OURS!)

In addition to helping out the bird population, keeping bird feeders provides humans with enjoyment and satisfaction, affording us intimate glimpses of, and connections to, the natural world in our own backyards. Bird feeding often serves as a "gateway" activity that leads people into deeper experiences with nature, benefiting themselves and often leading them into active conservation activities that benefit the birds in return.

as they produce eggs and bring food (mostly insects) to their young. At these times, birds may welcome the opportunity to visit a feeder for a quick meal. I do close down my feeders if local birds start bringing their fledglings, because growing birds require a lot of protein and calcium, which birdseed and suet just don't provide.

∨ ∨

Q I've been feeding birds for decades, but now I find myself on a pretty rigid budget and can't afford to spend so much. How can I keep feeding my chickadees and other favorite birds without going broke?

A Don't economize by buying cheap mixtures: you'll do more harm than good for the birds and won't attract as diverse an assortment of species. Your best bet is to buy sunflower seeds and a couple of very small window feeders (which will exclude pigeons, jays, and all but the most brazen squirrels). Fill them with a small amount of food at the same time each day. During times when there are more birds than the food can accommodate, they'll figure out alternative food sources, but your chickadees and their associates should quickly get into the habit of coming to feed at the same time each day.

In the long run, it's most economical to buy quality seed in large quantities. If you do this when you're feeding just a small amount each day, be sure to store your seed in a cool, dry place.

If You Build It, They Will Come

Q I set out my feeders a month ago and still haven't seen a single bird! What's wrong?

A It often takes a while for birds to discover a new feeding station, especially during the most severe periods of winter and the middle of summer. Local birds explore their area for new feeding opportunities during mild periods in winter. (That's why the busiest feeders have far fewer birds on nice days than during bad weather.) Once a few local birds discover a feeder, they'll draw the attention of others.

If you're still having trouble attracting birds, you may need to make the area around your feeders more enticing to birds.

˅ ˅

Q Where is the best place to put bird feeders?

A Place your bird feeders where you can enjoy the view, and where you can keep the birds safe. To minimize the chances that birds will collide with your windows, place feeders on poles or trees within three feet of a window. Feeders six feet or more from a window are the ones most often associated with fatal collisions.

(continued)

ATTRACTING BIRDS TO YOUR BACKYARD

Even the most attractive feeding station is no more than a fast food restaurant for birds without the other necessities that will make them want to stick around. Native plants that provide food, cover, and nest sites are your best choices. Locally native plants are perfectly adapted for the soil and water conditions of your area and perfectly suited to the needs of native local wildlife as well. You'll want an assortment of types, including nectar-producing flowering plants, fruit-producing shrubs and trees, and a variety of other trees for seeds, nuts, and nesting sites. Try to root out invasive weeds. Good sources of information about choices include local gardening and birding clubs, your county extension office, and your state or provincial department of natural resources.

A supply of clean water for drinking and bathing is also important. Many birds are attracted to birdbaths. Birds are especially drawn to the sound of dripping or flowing water, so setting up a plastic bottle with a small hole in the bottom above your birdbath to provide a slow, steady drip will bring in more individuals and more species than a birdbath alone.

Providing nest boxes and nest platforms will promote nesting. Fostering lichens and spiders will encourage tiny birds such as gnatcatchers and hummingbirds to nest nearby — they need bits of lichen and spider silk to construct their nests. You can also offer natural nesting materials such as cotton batting and short lengths of twine, placed in a clean suet cage or simply wedged into tree bark.

Feeders affixed to windows don't give birds enough distance to build up the kind of momentum that can kill or injure them if they do fly into the window. You can also place feeders directly on the window frame, or attach a window feeder to the glass with suction cups. As an extra measure of safety, select a location fairly close to trees or shrubs where birds can retreat when a predator approaches.

˅ ˅

Q How do I attract orioles to my feeders?

A Orioles enjoy nectar and fruit. You can attract them with sugar water in hummingbird feeders, but make sure the perches are large enough for them. Some manufacturers make sugar water feeders specifically for orioles. Orioles can also be attracted to grape jelly, although never offer this in amounts larger than a teaspoon — it's sticky! If you do offer jelly, it's best in small containers such as jar lids.

Orioles also come to fruits. You can offer orange and apple halves on platform feeders or by skewering them on manufactured feeders or on nails pounded into a small board. You can also try setting out fresh grapes or berries, or raisins and currants (softened by soaking first), in a cereal bowl or other small container. Make sure to throw away or compost any fruits that get moldy.

˅ ˅

A POND FOR THE BIRDS

Many people construct small backyard ponds to attract birds. These often attract insects, which will provide food for a broader selection of backyard birds. Make sure the water can't become stagnant or it will be a breeding nursery for mosquitoes. Once the pond is established, if toads and frogs lay their eggs in it, the tadpoles will help keep mosquito larvae in check, as will any young dragonflies and damselflies during their nymph stage before they emerge as adults. These insects are doubly helpful because the adults feed on adult mosquitoes.

Q I live in a high-rise apartment with a tiny balcony. Is there any way I can attract birds all the way up on the 17th floor?

A Depending on what the habitat below you is like, it may take some time for birds to discover your balcony. Bird feeders in high-rises along lakes and rivers are fairly likely to be discovered during migration. Feeders in any neighborhood are more likely to attract birds if there are trees and other vegetation at ground level, and the more plants on your balcony, the more likely curious birds will check it out. Providing food and nectar-producing plants may lure birds in, and will make your balcony more pleasant for you whether or not they ever arrive. You can learn more about attracting birds to city yards and balconies online at *www.celebrateurbanbirds.org.*

Q I love watching warblers and wish they would come to my feeders. Why don't they?

A Warblers are insectivores, not seed eaters. Some people offer mealworms at bird feeders, but most warblers instinctively search for insects in specific areas within specific kinds of trees, rather than searching in a spot that might lead them to a feeder. The time warblers are most likely to discover a mealworm or suet feeder is during migration, especially during harsh weather. When they're in an unfamiliar area, warblers often associate with chickadees, and a hungry warbler may notice the chickadees flitting back and forth from a feeder. After one warbler does discover a feeder, others may join it.

Pine Warblers and Yellow-rumped Warblers appear at suet feeders more often than most of their relatives, sometimes visiting a feeding station throughout a winter. Cape May Warblers, which drink nectar and feed on sap oozing out of holes drilled in trees by Yellow-bellied Sapsuckers, sometimes visit hummingbird feeders or an offering of sliced oranges.

Goldfinches are among the strictest vegetarians in the bird world, selecting an entirely vegetable diet and only inadvertently swallowing an occasional insect. Goldfinches even feed their nestlings regurgitated seeds, rather than the insects that most songbirds feed their young.

What's On the Menu?

Q Someone told me that bird seed mixes aren't good for birds — that I should be feeding them just sunflower seeds. But wouldn't mixtures give a more balanced diet?

A According to results of the Seed Preference Test conducted by the Cornell Lab of Ornithology, black-oil sunflower seeds attract the widest variety of species. Sunflower seeds have a high meat-to-shell ratio; they are high in fat (an important energy source for birds); and their small size and thin shell make them easy for small birds to handle and crack. (Striped sunflower seeds are larger and have thicker seed coats.)

Most mixtures include less expensive "filler" seeds such as millet and milo, which are often left behind as birds eat the good stuff. When these and other seeds in a mixture are left uneaten, they can start molding and can contaminate fresh seed. The Seed Preference Test showed that millet is popular with sparrows, blackbirds, pigeons, and doves but ignored by other birds; milo is preferred only by jays, pigeons, and doves. Nyjer seed attracts goldfinches, redpolls, and siskins.

Q I was told that it is bad to feed birds in the spring when they are feeding their young because the adults will feed the babies the sunflower seeds from the feeders rather than the more varied diet they should be feeding them. Is this true? Do I have to give up my feeders in the spring?

A That all depends. If goldfinches or siskins are bringing sunflowers to the babies, it won't hurt them — those nestlings or fledglings are pretty much vegetarians anyway. But once summer is underway, it might be bad to let cardinals, Rose-breasted Grosbeaks, or most other birds feed their babies too much sunflower seed or grape jelly, which don't have enough protein for growing babies. As long as the families are only coming to the feeders once or twice a day, they should be getting plenty of other proper food.

Sadly, a few bird parents are as clueless as a few human parents are about their responsibilities. A 2009 study by researchers at Binghamton University showed that some urban crows overfeed their young junk food, preferring items that were easier to obtain over more nutritious fare. Fortunately, most birds innately feed their chicks a natural, protein-rich diet.

Summer feeders are especially helpful when incubating birds need a quick bite before they need to return to the nest, and when adult birds are spending every waking hour searching out insects for their babies. A bird feeder can give them some quick energy to keep doing what needs to be done for their babies.

Q I was trying to buy some thistle seed to attract goldfinches, and the woman in the bird-feed store said they didn't sell it and wanted me to try something called "Nyjer seed." What is that, and how is it different from thistle seed?

A Nyjer seed, also spelled niger seed, is a tiny black seed that resembles thistle, and it's just as attractive to goldfinches, siskins, and redpolls. It can be fed in fabric tube feeders, called "thistle socks," or in tube feeders with tiny openings. It is heat sterilized to prevent germination. If a few weeds with yellow flowers do sprout up under your feeder, pull them before they go to seed.

ˇ ˇ

Q My goldfinches seem to love Nyjer seed, but after they've been feeding there is so much wasted seed on the ground! How can they possibly be getting any nutrition?

A What look like seeds on the ground beneath a feeder filled with Nyjer seed are mostly just the emptied-out outer shells. Finches slit open the outer coat and use their tongue to extract the tiny seed inside. Of course, the seeds are so tiny that when a finch pulls out one seed, a few others do spill on the ground. But finches usually arrive at feeders in flocks, and while some birds are sitting at the feeders, others are on the ground picking up the spilled seed.

13

Q How can suet be good for birds? Shouldn't we be concerned about their cholesterol levels?

A Most birds thrive on a diet high in fats and proteins. Unadulterated animal fat is actually good for them, especially during cold weather. Unlike humans, birds metabolize fat very efficiently. This gives them the energy they need to maintain their body temperature.

Never offer suet when temperatures get warm enough to make it goopy — it may coat the feathers, making them harder to preen and less effective at insulating the bird. If the bird is nesting, some of the soft fat may be transferred to eggs, plugging the tiny pores that provide oxygen to the embryonic chicks. Also, warm raw suet grows rancid rather quickly.

If you feed suet when temperatures are above freezing, it should be rendered, that is, cooked so that the impurities can

BANNING BACON

A great many birds love the taste of bacon fat. But like other processed meats, bacon contains nitrosamines, carcinogenic compounds formed from some of the preservatives used to cure the meat. Although nitrosamine levels in processed meats are far lower than they were in past decades, bacon does have detectable amounts of it. Compounding that, the very high cooking temperatures involved in frying bacon are conducive to nitrosamine formation. So despite the fact that birds love it, bacon and bacon fat pose too much of a risk to the long-term health of birds to warrant using it.

be strained off. Suet cakes sold in stores have been rendered and are usually okay for summer feeding during mild weather.

˅ ˅

Q I bought some suet cakes that were recalled because they'd been tainted with salmonella. Is this a common problem?

A No, it's not. Salmonella is unlikely to be a problem in rendered suet. In 2009, some suet cakes that contained peanut products were recalled because the peanuts had been contaminated during processing.

Peanuts and corn are exceptionally susceptible to contamination by bacteria and, especially, a fungus that produces extremely dangerous toxins, so those sold for human, pet, or livestock consumption are carefully screened. Screening is not required for products sold for wildlife consumption. For that reason, I prefer to buy suet cakes that don't contain peanuts or corn, although they are usually safe, and conscientious manufacturers do recall them if a problem is discovered. Whether you're feeding your birds pure, unadulterated fat or suet cakes made with other products, it's best to stop offering it if birds aren't finishing it quickly, especially in warm weather.

Red-tailed Hawks have been seen hunting as a pair, guarding opposite sides of the same tree to catch tree squirrels.

Q Is it okay to feed bread to birds?

A Bread gets moldy quickly, attracts rats and mice, and lacks the nutrients that most birds need. I strongly recommend other feeder offerings instead, especially sunflower seeds.

Ducks and pigeons can grow very fond of bread; it's especially important to leave bread off the bird-feeding menu in cities that prohibit the feeding of these birds.

˅ ˅

Q My neighbor puts out eggshells for birds to eat because she says it helps them get the calcium they need. Is that true?

A Yes. Especially during the nesting season, female birds require calcium to form strong eggshells. Birds can get calcium from natural foods, such as small snails, sow bugs, and slugs. But recent studies have found that in some areas acid rain leaches calcium out of the soil, possibly making it harder for these prey species to get enough calcium, ultimately affecting the birds.

You can provide a good source of calcium for the birds by crushing shells from hard-boiled eggs and setting them out for the birds to eat. If you use shells that haven't been cooked, bake them in the oven at 250 degrees for 20 minutes to protect birds from salmonella. Don't microwave them — they may shatter!

When birds are suffering from calcium deficiencies, they sometimes eat inappropriate substitutes. Blue Jays have been observed chipping and consuming house paint, especially in the Northeast when snow is covering the ground. Researchers believe the Blue Jays are interested in the calcium found in paint and that they are stockpiling the paint chips for spring. Unfortunately, paint also contains ingredients that might not be so healthy for the birds. Providing eggshells can help them while preserving your house!

WEDDING RICE — BAD FOR BIRDS?

Rice absorbs a lot of water when cooked, so some people worry that it might swell in birds' stomachs, causing their stomachs to explode. But many birds eat uncooked rice in the wild. Bobolinks are even nicknamed ricebirds for this food preference. Even though rice thrown at weddings won't result in exploded birds, some places ban throwing rice or birdseed at weddings for other reasons: slippery seed can be a hazard for wedding guests wearing smooth-soled dress shoes, and leftover seed or rice on the ground may attract mice.

Q **Is it okay to feed peanuts to birds? Should they be salted or unsalted?**

A You can feed birds unsalted peanuts, but beware of what you buy. Peanuts can harbor dangerous toxins produced by either of two harmful fungi, *Aspergillus parasiticus* or *Aspergillus flavus*. These aflatoxins can be lethal for humans and livestock, so peanuts sold for human consumption or for livestock feed must be screened for them. But there are no similar regulations regarding peanuts sold for feeding wild birds, so these dangerous poisons may be present in peanuts sold at bird-feed stores.

Peanuts sold in grocery stores have been screened. If you buy peanuts specifically sold as bird feed, make sure they're clearly labeled as free of aflatoxins. Always store peanuts (for yourself as well as for birds) where they will stay completely dry, since moisture encourages fungal growth.

I would not use salted nuts. There are no indications that any birds prefer salted over unsalted, and the salt may be harmful.

˅ ˅

Q **Does peanut butter stick to birds' mouths so they can't eat, and they eventually starve?**

A I know of no studies that support this, but it seems prudent, especially in warmer weather, to mix peanut butter with something gritty such as crushed eggshells or cornmeal. If the weather is warm enough that peanut butter gets goopy, bring it in. Softened peanut butter can stick to belly feathers, reducing their waterproofing and insulation; and worse, if the bird is incubating eggs or brooding young, those sticky feathers may smear the oil onto eggs or baby birds.

ˇ ˇ

Q **What are mealworms and is it good to feed them to wild birds?**

A Mealworms are larvae of a flightless beetle of the species *Tenebrio molitor*, and many birds love them. Mealworms are probably the single most inviting feeder offering for blue-birds, and if tanagers or warblers discover them, they may come right to your window to feed.

You can purchase mealworms in small quantities from bird-feed stores and bait shops, or in larger quantities from mail order distributors. Mealworms can be a pest insect in granaries, but in captivity they are easy to keep in small containers such as ice cream buckets or plastic tubs.

Feed the mealworms oatmeal and grains along with a few small pieces of potato, carrots, or apple as a water source. If mealworms arrive packaged in wadded newspaper, transfer them into buckets with proper food as soon as possible. They

eat paper, and the toxic inks that may be present on the paper would be swallowed, which is probably bad for birds. In a cool basement, mealworms can be kept alive without reaching their adult stage for weeks. (Birds seem to prefer eating the larval-stage grubs to the pupae or adult beetles.)

Offer mealworms in small bowls or small acrylic bird feeders. (Tape any drainage holes closed to prevent the grubs from squeezing out.) Once birds discover them, you may be inundated. Your best strategy is to set out a small handful at one or more specific times of day. If you whistle every time you set them out, your chickadees may soon start flying in the moment they hear you.

Adaptations for Eating

Q Great Blue Herons, loons, eagles, and puffins all specialize on catching and eating fish, but they look so different! I thought things that caught and ate the same food evolved to look similar.

A All of these species do eat fish, but each has evolved a different way of capturing them and feeding them to their young. Herons stab or grab their fish from a standing position. Their bill and neck muscles are powerful and their feet are wide to support their bodies in wet mud without sinking. They catch most of their food in shallower water than do loons, which chase their fish through deeper water. Like herons, loons grab or stab a fish with their bills. In both cases, the feet and bill are well adapted for catching fish but are useless for carrying or tearing fish apart, so they swallow a fish whole.

Herons nest in trees, and their nestlings remain in the nest for several weeks after hatching. To feed them, herons first swallow the fish catch. With the weight in their stomachs, close to their center of gravity, they fly back to the nest to regurgitate the semidigested fish right into their nestlings' mouths. Loon chicks follow their parents, who give them tiny fish and large aquatic insects until the babies learn to catch them on their own. They don't need to transport fish anywhere.

When puffins spy a school of fish from the air, they dive into the water and start snapping up the little fish. They swallow some under water, or when the fish are abundant, may catch several to eat at the surface. Puffins nest in burrows. They feed their young whole little fish, which they carry back to land in their bills. The roofs of puffin mouths and their thick, muscular tongues have backward-facing spines that help them hold fish in their mouths even as they're catching more.

When fish are abundant, puffins catch them close to their nesting sites, but because they may have to travel as much as 80 miles, or even more when fishing is poor, the adaptations that allow them to carry many fish at once help reduce the number of trips they must make to the nest. They can hold a dozen or more fish in their bills at one time. It's a myth that they line up the fish with the heads facing alternate ways.

Bald Eagles and Osprey catch their fish with their feet, which are conveniently located near their center of gravity, so they can easily carry the entire fish back to their nests to feed their young. These raptors have sharp, hooked bills designed for tearing apart fish, so they eat, and then feed their young, chunks of fish rather than swallowing them whole.

ˇ ˇ

Q **Which birds especially like to eat bees?**

A Bees are dangerous! Unless a bird knows instinctively how to safely catch and eat bees, either by selecting drones (which have no stingers) or by removing the stinger before swallowing its prey, it can be harmed or killed. I once witnessed a year-old captive American Crow catching a wasp in midair. The wasp stung its upper throat, and the bird died within minutes. Bees are loaded with nutrients and are fairly large, however, so it's not surprising that some birds would have adapted to overcome their powerful defenses.

In North America, two birds are so good at catching bees that both are nicknamed the "bee bird." One is the Eastern Kingbird. One study found that more than 32 percent of an Eastern Kingbird's summer diet is composed of bees, ants, and wasps. Western Kingbirds also eat a great many bees.

The other is the Summer Tanager, which is often found near apiaries. A Summer Tanager snaps bees in its bill (which is longer than most tanager bills, perhaps to hold these danger-ous insects farther from its face) and carries the bee to a perch where it first slams it repeatedly against the perch to kill it and then removes the stinger by wiping the dead insect on the branch.

Beyond North America, one bird family, the bee-eaters (Meropidae, in the same order as our kingfishers), specializes on catching bees. Bee-eaters, found in Africa, southern Europe, southern Asia, Australia, and New Guinea, are beautifully colored birds. They catch bees in their long bill, and then, like our Summer Tanager, swallow the bee only after removing the stinger.

˅ ˅

Q **I often see flickers in my yard picking at the ground. Do they eat ticks?**

A No, flickers specialize on ants. When a flicker finds an anthill, its long, sticky tongue traces the deep tunnels and pulls out lots of delicious ants. Of course, ants aren't so delicious to most birds — their bodies contain formic acid (their family name, Formicidae, highlights this fact). In the 1800s, when many people bought birds at town markets to consume, some relished the spicy flavor of flickers while others abhorred it. John James Audubon, who tasted every bird he painted, wrote of the flicker, "I look upon the flesh as very disagreeable, it having a strong flavour of ants."

˅ ˅

A CHANGE OF TASTE

American Goldfinches spend their entire winter eating sunflower and Nyjer seeds at feeders, and dried weed seeds in the wild. In spring, when the first dandelions go to seed, and as other natural food sources become available, goldfinches begin spending most of their time eating far from feeders. It's very disappointing for those of us who watch feeder birds, because this usually happens right as the goldfinches are coming into their most gorgeous plumage.

Although goldfinches start pairing in April, they do not begin nesting until much later. Indeed, the American Goldfinch is one of the latest nesters in North America. In the East, goldfinches normally wait to begin nesting until late June or early July, when thistles, milkweed, and other plants with fibrous seeds have gone to seed. They use this downy material in nest construction, and the seeds themselves provide food for the chicks.

Q We've noticed that every year around mid-April, many of our backyard birds seem to disappear for about two weeks. Where do they go or what is happening?

A In April, many birds that spent the winter with us have migrated farther north. You won't see them at your feeders again until late fall. At the same time, as natural foods, usually including insects, become more abundant, many year-round residents lose interest in feeders. Also, many resident birds gravitate to the areas where they'll nest to feed, so they can spend more time courting and defending their territories.

Meanwhile, depending on weather, some birds that wintered in the southern states, such as White-throated Sparrows, haven't yet arrived at their summer places, and Neotropical migrants are still even farther south. During the lull, we get a bit of time to savor robins, bluebirds, phoebes, and other early migrants before the floodgates of May open.

ᴧᴠ

Bird Behavior at the Snack Bar

Q I've been a casual bird-watcher for many years but only recently started feeding them, and it's opened a whole new world to me! It's such fun going beyond just identifying them to observing their behavior. But now that I'm paying attention, I'm wondering why my chickadees come in flocks the way my goldfinches do, but the chickadees never sit side by side and eat together the way the finches do. Why is that?

A I agree that watching bird behavior is fascinating! Every species has its own set of behaviors that make it successful in its own habitat.

Goldfinches are one of the rare species that eat almost entirely seeds, even feeding their nestlings mostly a seed diet rather than the insects that most growing young songbirds need for protein. Goldfinches and their relatives tend to eat food items that are "patchy" — abundant in a few spots at a time while not at all available in other places. By spending time in a flock and wandering widely, they have plenty of opportunities to discover new patches of appropriate food. But these food supplies can be depleted within a short window of time, either as goldfinches or other animals eat them or as the plant sheds them. This is factored into the game plan of nomadic flocking birds: when the food is used up, they move on. So whether in the wild or at your feeders, normal goldfinch behavior is to feed together.

Chickadees aren't nomadic — a winter flock stays within an area of about 25 acres (10 ha) for the entire season. Chickadee flocks benefit from many eyes to spot predators and new food resources, but individual birds hide and store seeds and other items for later feeding, so individuals prefer to keep more distance between themselves and other flock members. Usually chickadees space themselves from about 3 to 30 feet (1–10 m) apart, whereas the distance between goldfinches at a feeder can be measured in inches.

PUTTING YOUR OBSERVATIONS TO WORK

If you enjoy feeding birds, consider joining Project FeederWatch at *www.feederwatch.org* and contributing observations of all your feeder birds. Your sightings help scientists understand which changes at your feeders represent declines of birds across large areas, and which ones may reflect normal changes in the abundance of birds from one season to the next.

Q My grandmother in upstate New York has had bird feeders for as long as I can remember. The first Evening Grosbeaks I ever saw were swarming all over her feeder. Now she says that the grosbeaks hardly ever come to her feeders any more, and when they do, there are only a few. Are they becoming rarer than they used to be?

A Yes, unfortunately. The decline has been documented by participants in Project FeederWatch, which asks bird-watchers to count and report the number of birds they see at their feeders during winter. FeederWatch data gathered between 1988 and 2006 showed a 50 percent decline in the percentage of sites reporting Evening Grosbeaks. At locations where the grosbeaks were still being seen, average flock size had decreased by 27 percent. The cause is still unknown, but data from FeederWatch are bringing more attention to this worrisome decline and may eventually help us understand why it's happening.

Q I went on a local field trip and the leader said chickadees eat a lot of insects in the winter. How is that possible?

A Because insects are cold-blooded, they may not be active in winter — but they're out there! Many spend the winter months as eggs or pupae hidden in crevices in bark and twigs. Black-capped Chickadees are masters at locating them with their sharp eyes and at grabbing them with their tiny beaks, which can reach into crevices to pull them out. Chickadees also occasionally chip frozen fat or meat from carcasses. This shouldn't be surprising considering they also feed on suet.

When someone offers mealworms, or a chickadee discovers an abundant source of insects, it may start storing, or *caching* its prey, but normally chickadees eat insects as they discover them in winter, caching seeds instead.

Although primarily insectivorous, the Black Phoebe occasionally catches fish. It dives into ponds to catch small minnows or other tiny fish and may even feed fish to its nestlings.

How to Keep a Hummingbird Happy

Q **What's the best thing to feed hummingbirds — honey or sugar? And should I use food coloring?**

A Use a sugar solution, leave out the food coloring, and never use honey. Honey is more natural than processed sugar, and so some people assume it's more nutritious. But honey fosters rapid bacterial and fungal growth. Always use regular processed sugar in your hummingbird feeders.

Hummingbirds are attracted to the color red, so some people add red food coloring to the sugar solution. Food coloring adds no nutritional value, however, and is harmful to hummingbirds. Attract the hummingbirds by choosing a feeder with bright red parts.

v v v v v v v v v v v v v v v v v v v v v v v v v v v

SUGAR SOLUTION FOR HUMMERS

🐦 **To make a sugar solution,** mix ¼ cup of sugar per cup of water, a good ratio especially during hot, dry conditions when hummingbirds may be somewhat dehydrated. During cold, wet periods you can strengthen the mixture, especially during spring and fall migrations, to ⅓ cup of sugar per cup of water. Boiling isn't necessary if you use clean containers for measuring, use the sugar solution immediately, and change the solution every two or three days. If you mix up larger batches to be stored in the refrigerator, boiling is a good idea.

Q My hummingbird feeders keep attracting wasps and ants. What should I do?

A Ants can easily be controlled at suspended hummingbird feeders that are designed with a central moat at the bottom of the suspension wire rod. Fill the moat with plain water and ants won't be able to reach the sugar water at all.

Unfortunately, there isn't such a simple solution for discouraging bees and wasps. I'm not sure why so many feeders are equipped with yellow bee guards because bees and wasps are attracted to the color yellow, so the bee guards actually attract their attention. If solution drips from the feeder ports onto the guards, the bees get a tasty meal. I would never set out wasp traps or put anything toxic, such as insect repellents, on feeder ports — the risk of contaminating the food is too high.

I learned of one effective way of dealing with bees from an Oklahoma birder named Phil Floyd, whose wife figured out how to lure the wasps away from the hummingbird feeders without harming them. Phil wrote, "She filled another feeder with extra sugar content and also sprayed the outside of that feeder with the stronger sugar solution. She hung it in an area away from the other feeders. Then she went to the hummingbird feeders covered with bees, scraped several of them off into a jar, and took them to the new target feeder. She repeated this a couple of times until word among the bees spread and they all started going to the new feeder. We continued to keep the 'bee feeder' full of extra sugary water and kept spraying the outside with that same sugar solution. The bees took only to that one. Problem solved!" The heavy bee activity at the extra-sugary feeder discouraged hummingbirds from visiting that feeder.

⌄ ⌄

Q **I've heard that we should bring in our hummingbird feeders before Labor Day; otherwise hummingbirds might stay at the feeders instead of migrating. Is that true?**

A No. Like other migratory birds, hummingbirds grow restless as day length decreases in fall, and not even the most wonderful feeding station can hold a hummingbird when the urge to migrate kicks in.

Stragglers aren't stupid or lazy birds being enticed to remain by our feeders. Their bodies just weren't quite fat and muscled enough to start the journey when the others departed. A few late Ruby-throated Hummingbirds are adult females whose first nest failed. It takes time for hummingbirds to recover after raising young, and our feeders can help them get enough calories to fatten up, which is especially important as flowers start drying up. Most late ruby-throats are juveniles that haven't yet fully bulked up before they can leave. Our feeders can make the difference between life and death for some of these, especially after the first frosts.

˅ ˅

Q **The hummingbirds at my feeder seem to fight all the time. My feeder has eight ports — why can't they share?**

A Hummingbirds are aggressive for a good reason: they can't afford to share flowers during times when not many blossoms are available. They may have to wander a long way after nectar is depleted. This aggression is so deeply ingrained that they just can't figure out that the "nectar" in feeders virtually never runs out and doesn't really need to be defended.

Overall, you'll feed far more hummingbirds by setting out four tiny one-port feeders, spread out so a bird at one doesn't easily see the others. You'll get to watch hummingbirds through more windows, and they'll be much happier, too.

RARE VISITORS: VAGRANT HUMMINGBIRDS

If you keep your hummingbird feeders up during the fall anywhere in the United States and southern Canada, you may get a surprise visitor. Hummingbirds from the tropics may suddenly appear, even as late as December. These poor birds are often doomed, but since our feeders are their only chance of survival, we can add days, weeks, or months to their lives. Some may even survive long-term, eventually heading back to more appropriate areas.

Rufous Hummingbirds breed in the northern Rocky Mountains and usually winter in Mexico and the Southwest. Some individuals winter in the southeastern states, and a handful even spend early winter in central and northeastern states.

Q My hummingbirds seem to be acting okay, but a yellowish powdery substance seems to be covering their faces and the base of their bills. I once read that hummers get a fungal disease when people feed them honey. Is this what's happened to my birds?

A Honey does foster mold and bacteria growth, so don't offer honey in a feeder. But from your description, your birds don't sound ill at all — hummingbirds have such a high metabolic rate that when they get an infection, they get obviously sick and die pretty quickly. The yellowish powder on their faces and bills is probably pollen from the flowers they've been feeding on.

Mourning Doves usually feed on the ground, swallowing seeds and storing them in an enlargement of the esophagus called the "crop." Once they've filled it (the record is 17,200 bluegrass seeds in a single crop!), they can fly to a safe perch to digest the meal.

Coping with Unwanted Guests

Q **So many grackles are pigging out at my feeders that the little birds can't get in. What should I do?**

A Grackles and most other blackbirds have bills that aren't designed to crack open seeds with thick shells. Try offering striped sunflower seeds only, rather than seed mixtures or black oil sunflower seeds. Black oil sunflower is a little more nutritious but has a much thinner shell, easier for grackles and also House Sparrows to open. If you have a platform feeder, you can also try switching to a tube feeder, since grackles don't usually visit hanging feeders.

Switching to striped sunflower seeds may also deter other birds that may be unwelcome feeder visitors, such as European Starlings and House Sparrows. These invasive species were brought to America from Europe. They nest in cavities but cannot excavate them themselves, so they take them from woodpeckers, bluebirds, and other vulnerable native species. It's best to avoid subsidizing starlings and House Sparrows if you can.

GOOD-FOR-SOMETHING GRACKLES

Great-tailed Grackles are abundant and noisy, so they are generally considered nuisances in their range. But I will always have a soft spot for them. One morning when I was birding in Las Vegas with my first baby, I set him in his car seat on a picnic table in a city park while I scanned nearby trees. The Great-tailed Grackles were courting, which means they were making loud, bizarre whistles, rattles, and a variety of other cool noises. Joey giggled and cooed, happily engaged and entertained for more than 45 minutes while I looked at a variety of other birds I'd never have been able to enjoy but for those grackles.

Q My friend has a 40-foot pond with ducks in a rural setting in Arizona. She has a problem with a large population of grackles. Their noise and mess are driving her nuts. She wants to know what to do to get them to leave the area.

A Remember the line from the movie *Field of Dreams*, "If you build it, they will come"? In the case of Great-tailed Grackles in Arizona, if you provide water, they will come. These birds are so well adapted to urban, suburban, and rural areas with human habitation that once they colonize an area, they are very difficult to disperse. No matter what your friend does to discourage them, she may lose all her other birds before the grackles take the hint.

Tell her to make sure she's not subsidizing them with food (all her feeders should be a type that excludes large birds), but if there are good supplies of natural food, she may simply be stuck. A simple solution if she happens to have a dog who chases birds and the time and patience to be vigilant for a few weeks: let the dog out to scare off the grackles whenever they make an appearance.

˅ ˅

Q Can you name one good thing about pigeons?

A Pigeons came to America with early settlers, who brought them from Britain and Europe for food and sport. Pigeons have saved human lives by carrying critical messages during emergencies and wartime. One English hospital uses pigeons to transport blood samples to its laboratory across town, saving money and avoiding traffic jams.

Charles Darwin watched pigeons closely as well as raising his own, honing his theories by studying how different forms evolved due to selective breeding. Psychologist B.F. Skinner conducted many experiments on pigeons to study learning, and claimed that they were extremely intelligent animals.

At feeders, pigeons may be large and sometimes numerous, but they're not particularly aggressive toward other birds. They don't compete with native American birds for food or nest sites, and they do provide food for urban Peregrine Falcons.

STUMPING THE SQUIRRELS

Many people who feed birds wonder how to attract them without bringing in squirrels. Whole industries appear to have sprung up around creating the perfect way to foil squirrels, but often that tiny rodent brain seems more than a match for the average engineer.

My father-in-law designed an effective squirrel baffle using a large pizza pan. He cut a hole in the center, just a little larger in diameter than his feeder pole. He wrapped electrical tape around and around the pole where he wanted the baffle and then placed the pizza pan above the tape. Squirrels could not climb up the pole with the pan blocking the way, and couldn't leap on the pan because it was so floppy they couldn't get a footing. A cone of aluminum sheeting can also keep squirrels from climbing up the pole. Commercial baffles are also available.

But baffles work only if the feeder is too high and too far from trees for squirrels to leap to directly. If your feeder is anywhere within 8 feet (2.4 m) or so of a tree, a squirrel is eventually going to jump across and then teach its friends. Feeders on windows are also fair game. If there is a tree near your house, any self-respecting squirrel will quickly figure out how to leap to the roof and drop to the feeder from there. I've even seen large gray squirrels squeeze into my tiny acrylic window feeders.

Another option is to purchase a feeder that has a cage around it, allowing small birds, but not squirrels, to get in and reach the seeds. These feeders may not be as aesthetically pleasing, but they will prevent squirrels from demolishing the seed supply. Of course, there is one more option — learn to enjoy squirrels and their antics.

> A pigeonhole is a small, open compartment (as in a desk or cabinet) for keeping letters or documents; when we "pigeonhole" people or things, we place them in categories, usually failing to reflect their actual complexities. The term comes from the small nest compartments in pigeon coops.

Q It's illegal to feed pigeons where I live. How can I keep them out of my bird feeders so I don't have to pay a fine?

A If pigeons are coming regularly to your feeders, it may take some time and ingenuity to shoo them elsewhere. First and foremost, make sure seed doesn't collect on the ground beneath your feeders. Also, pigeons are far more likely to come to platform feeders than any other kind, so you may want to close down those feeders and offer alternatives for the duration. Try a hanging feeder, since pigeons virtually never come to hanging feeders of any kind.

Housekeeping for the Birds

Q What's the best way to clean a feeder?

A Hold it under running water in a laundry tub or with a hose in the yard while scrubbing with a brush. If you do this every few weeks and there is no evidence of illness in birds in your yard, you don't need to do anything more.

Beyond keeping your feeders clean, it's a good idea to rake up old seeds and shells beneath your feeders every few weeks, and more often during wet spells.

∨ ∨

Q How often should I clean my birdbath?

A Clean your birdbath whenever it is visibly dirty, and at the very least every four or five days to prevent any mosquito eggs from hatching and the larvae from emerging. This is important not only because mosquitoes are pests to humans but also because they can spread disease, including West Nile virus. The best way to deal with algae without harming birds or our environment is to be proactive, scrubbing at the first sign of algae and rinsing thoroughly rather than resorting to detergents or bleach. Use clean water and a stiff scrub brush.

BIRDS DON'T NEED HOT TUBS

I'm sometimes asked about providing heated birdbaths in the winter. Generally, this isn't a good idea. All animals require drinking water, and birds are no exception. They're drawn to open water, and find birdbaths especially attractive when natural sources of open water are scarce.

But when temperatures are well below freezing, open water steams up, and birds visiting it may become coated with ice. Worse, some birds may be tempted to bathe rather than simply drink, and when they hop out of the water, ice may form on their feathers, making flight impossible. I've heard firsthand accounts of this happening to European Starlings and to Mourning Doves.

I would never use a heated bath when temperatures were below about 20°F (–7°C) to prevent steam from coating feathers. If using a heated bath, it's also a good idea to cover it with a grille of wooden dowels that allow birds to insert their beaks for drinking without being able to get their bodies into the water.

One thing to consider before buying a heated birdbath is whether it's worth expending the natural resources to run the electricity. I often set a sturdy plastic bowl filled with water on my window platform feeder in the morning, and I bring it in when ice forms. Thirsty birds can get a drink, but since they can also get the water they need from snow and dripping icicles, I don't feel that they're deprived.

Q **My neighbor says bird feeders are dangerous for birds because they foster the spread of diseases. Is this true?**

A When birds are extremely concentrated, in either natural situations or at feeding stations, sick individuals can spread their germs to the other birds, so it's prudent to close down your feeding station if you spot any sick birds. Wait several days after seeing the sick bird to replace the feeders, and thoroughly clean them and set out new seed when you do.

It's also important to keep the ground beneath your feeders raked and clean of old, moldy seeds. When seeds collect on the ground beneath feeders and start to rot, birds may pick them up in their mouths. Bacteria growing on rotten seeds may spread to clean seed. Birds may get sick at one feeder and then spread their illnesses to other feeding stations through their droppings.

Although it's possible for birds to become sick at feeders, I don't want to overstate the potential harm. Our children may be exposed to dangerous germs at school, but overall the advantages far outweigh this small possibility, and we trust that conscientious parents will keep sick children home to minimize the danger. If there's a risk of a serious epidemic, schools are closed. In the same way, the advantages of feeders usually outweigh the dangers as long as we're proactive about cleanliness and quick to react when a sick bird does show up.

SEE ALSO: *chapter 3 for more on dealing with sick birds.*

Q **Can bird feeders give people diseases?**

A Public health departments don't consider bird feeding a risk to humans. The vast majority of bird diseases don't affect humans and the few that can, such as botulism and salmonella, are virtually never passed on to humans through contact with feeders. However, you should take a few easy and common-sense steps to ensure that bird feeders are safe for you and the birds.

Washing your hands after handling feeders and seeds is always prudent. You should also keep feeding stations clean and rake the area beneath feeders. Because the bacterium that causes botulism is found in soil regardless of the presence of bird feeders, it's also prudent to wash hands after raking leaves or doing other yard work.

If you discover a sick bird in your yard or a bird that may have died of illness, close down your feeding station for at least a couple of weeks to prevent disease from spreading to other birds.

⌄ ⌄

Q **My landlord says he doesn't want any bird feeders in our apartment complex. He claims that bird feeders attract pigeons, rats, and other vermin. Is this reasonable?**

A Pigeons, rats, mice, and other animals that associate with humans are always searching for food resources, so if your feeders or spilled seed are accessible to them, they'll come. Because rats and mice pose serious human health issues in urban areas and are difficult to control, it's very important to be proactive.

Choose feeders on poles that rats and mice can't climb, with wire mesh or weighted perches that exclude larger birds and squirrels. A seed catcher beneath to prevent seed from spilling on the ground is also a good idea. Keep your feeders meticulously clean, and clean up any spilled seed and hulls daily. Sometimes landlords are concerned about general sanitation, since rotting seed hulls building up near a building's foundation can cause structural problems. If you can assure your landlord that you take his concerns seriously and are willing to do the work necessary to prevent problems, you may be able to change his mind.

˅ ˅

Q I have wasps in my birdhouse. What should I do?

A Wasps and bees seldom usurp boxes from nesting birds. They are mostly found in empty boxes. If these insects are found in a box, it is best to let them be and not take any active measures to exterminate them. Instead, wait to clean them out in the fall when the weather is cooler and their activity has halted. You can prevent wasps and bees from establishing themselves by applying a thin layer of soap (use bar soap) onto the inside surface of the roof. This will create a slippery surface between the insects and the roof of the box. Never use insecticides in any nest box.

CHAPTER 2

Birding for Fun and Science: The Sport of Bird-Watching

Joseph Hickey once defined bird-watching as a disease "which can be cured only by rising at dawn and sitting in a bog." That may not be quite true, but why *do* people watch birds? And once our interest is sparked, what equipment do we need, and how do we go about finding and identifying birds?

Every year I field hundreds of questions about how to watch birds. What are the best binoculars? How do I pick a field guide? Is it best to go birding alone or with a group? I love helping people start bird-watching. After all, no one should go through life listlessly.

Using Binoculars Like a Pro

Q I'm interested in doing more than just looking at my backyard birds, but isn't birding an expensive hobby, with state-of-the-art equipment and a lot of travel?

A Birding doesn't have to be expensive, though it certainly can be for those who purchase the best optics, the most current electronic gadgets, and the airplane tickets to embark on world travel. But it can be equally satisfying, and sometimes even more so, to watch birds while spending very little money. You can have years of enjoyment with excellent binoculars costing less than $300 that will allow you to see and identify as many birds as those top-of-the-line ones. Investing $30 in a field guide can provide a lifelong reference for learning about hundreds of birds in your own area and anywhere else you may go in North America. And birding locally can provide endless enjoyment and excitement as you hone your skills and continually learn more about the diversity and behavior of birds.

˅ ˅

Q **My husband bought me a really great pair of binoculars, but whenever I try to look through them, everything sort of blacks out and I can't see a thing!**

A Considering how expensive binoculars can be, it's odd that most companies don't include operating instructions in the package. Using binoculars is like riding a bike — wonderfully easy, once you have the hang of it.

Before you try to see birds through your binoculars, you need to make a few adjustments. Virtually all binoculars have several helpful features that allow them to be tailored to different users. The eyecups hold the ocular lenses (the lenses you look through) exactly the right distance from your eyes (this distance is called *eye relief*), to optimize magnification and cut out peripheral light, making the image clearer and brighter. Extend the eyecups if you don't wear eyeglasses. Since eyeglasses hold binoculars away from the eyes and let in peripheral light anyway, retract the eyecups if you do wear glasses.

Next, set the barrels of the binoculars to match the distance between your eyes. Looking through them, adjust the barrels until you have a solid image through both eyes. If the width isn't set properly, your image will black out.

Virtually all binoculars on the market have center focusing, in which a single knob or lever controls the focus for both eyepieces simultaneously. Our eyes are seldom precisely matched, so to accommodate the difference between our two eyes, binoculars also have a diopter adjustment near the optical lens on one side or the other, or as part of the center focus knob. Diopter adjustments are normally numbered from +2 to –2.

Here's how to adjust your binoculars so you can use them without eyestrain:

▶ First find the diopter adjustment and set it at zero.

▶ Find something a good distance away that has clean lines. A sign or something else with letters or numbers is often a good choice.

▶ Cover the objective lens (the large outside lens of the binoculars) with the lens cap or your hand on the side controlled by the diopter adjustment, and then focus on the sign using the center focus knob. Try to keep both eyes open as you do this.

▶ Switch hands, uncovering the lens with the diopter adjustment and covering the other lens. Focus again, this time using the diopter adjustment, not the center focus.

▶ Repeat each adjustment a couple of times. After you're done, your sign should be crisply focused through both eyes.

▶ Notice the number setting on the diopter adjustment. Sometimes during normal use, the adjustment knob may get shifted, so every now and then, check to make sure it's set where it should be for your eyes.

▶ Finally, make the neck strap as short as it can be while
still allowing you to use the binoculars comfortably and
put them over your head easily. The longer the strap is,
the more the binoculars will bounce, and the greater the
chance you may bonk them against rocks, tables, and
other objects whenever you bend down.

˅ ˅

Q I think I have my binoculars perfectly set for my eyes,
but I just can't find any birds with them! There can
be a cardinal sitting in a tree right in front of me, but when
I pull up the binoculars and try to scan, the bird flies away
before I can find it. What should I do?

A When you find a bird, keep your eyes on it while you
place the binoculars in front of your eyes and focus without moving your head. Practice this with inanimate objects
first. Starting with large items, test yourself to see how quickly
you can spot something and pull up your glasses to view it. It
can be frustrating at first, but once you master finding inanimate objects and resting birds, finding moving birds through
your binoculars will soon become second nature.

> The largest Red-tailed Hawks weigh three pounds.
> A dog about the size of a red-tail would weigh at
> least 10 times that!

BUY THE BEST BINOCULARS YOU CAN

There are hundreds of kinds of binoculars, and almost as many opinions about how to choose them as there are birders. The trick is to find the ones that are best for you and your budget, keeping in mind six key features of binoculars: prism type, magnification, brightness, field-of-view, comfort, and price.

Types of Binoculars

There are two basic binocular types: roof prisms and porroprisms. There is one more mirror inside roof prisms, which means the view is slightly dimmer, which is a serious consideration when you're buying on a budget. For the same price and manufacturer, porroprisms give a slightly better image. That said, the design of porroprisms makes it easier for dirt and grit to get inside the binoculars, and they're harder to waterproof. Most top-of-the line binoculars are roof prisms because they are so much easier to keep clean. Avoid choosing binoculars that have separate focus adjustments for each barrel — these are a very poor choice for birding, when you need to locate the bird and focus quickly.

Magnification and Brightness

Every pair of binoculars is described with two numbers, such as 8×40, 10×50, or 6×32. The first number is the magnification and the second number tells you, in millimeters, the diameter of the objective lens, which affects the brightness of the image.
The magnification number tells you how much closer objects appear through the binoculars. Ten-power glasses bring things ten times closer. For otherwise identical binoculars, the higher

the power, the closer the bird appears. There is a trade-off: the image will appear dimmer and the field-of-view will be smaller than through the same binoculars with lower magnification.

I use 6x binoculars, and if I'm looking at the same kettle of hawks as a birder next to me with 10× binoculars, I will be able to see more birds in my field of view at any given moment. If some-one yells out that there's a Golden Eagle flying in, I'll usually be able to find it more quickly because of that wider field of view. The trade-off is that the other birder will be able to see more detail on the birds because of the greater magnification.

If there's a strong northwest wind and we're both shivering, the view through my 6x binoculars will be less shaky. If we're looking at shorebirds on a hot, sunny afternoon, my 6× glasses will show less distortion. For a given price point from the same manufacturer, the optics of lower-power glasses virtually always provide a crisper, clearer view than those of higher-power glasses, especially noticeable with lower-quality binoculars. And yet, those higher-power glasses do bring the birds closer.

For binoculars of the same magnification, the bigger the second number is, the brighter your image — a good thing. But the bigger that number is, the heavier the binoculars are — a bad thing. You'll optimize the amount of light gathering if the second number is at least 5 times the first number; that is, 6×30, 7×35, 8×40, or 10×50. This guideline is particularly important when you're buying inexpensive or even mid-range binoculars, which don't have the superior coatings and special low-dispersion glass that top-of-the-line binoculars have.

Comfort and Ease of Use

This should be another important consideration. Don't choose binoculars that are so heavy you won't want to use them. Make sure that when you look through them, you see one full image. If the eyepieces are too far apart, or if the binoculars aren't well suited to the kind of glasses you wear, you may see two images or a partial image. It should also be easy for you to reach and turn the focus wheel.

Try binoculars within the highest price range you can comfortably afford, and rank them based on your preferences for magnification, brightness, sharpness, field of view, and comfort. Try to look at the same object in the distance with each pair so that you get a useful comparison. Once you settle on a pair of binoculars, stop reading about new models and comparing your glasses with those of other people! There really is no perfect binocular, and you'll maximize your satisfaction by putting your focus back on birds.

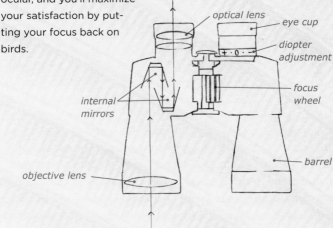

optical lens

eye cup

diopter adjustment

focus wheel

internal mirrors

barrel

objective lens

Q I share my binoculars with my wife, and our eyes are different. Do we need to shift the diopter adjustment every single time we pass the binoculars between us?

A Although you'd both get your clearest view by readjusting the diopter every time, that's sure not a fun way to go birding! You might try setting the adjustment halfway between her number and yours. If this bothers one of you more than the other, shift it a little toward that person's proper setting. Some married birders regret not checking out their partner's diopter adjustment number before making a lifetime commitment. Others just go ahead and buy a second pair of binoculars.

⌄ ⌄

Q Two experienced birders in my town got into an argument on a field trip, debating how often they should clean binoculars. One said he cleans them before every birding outing. The other said you should never clean binoculars unless they're noticeably dirty. What do you think?

A Most of the value of expensive binoculars is due to the quality of the lenses and coatings. Sand, dust, and other particles on the lenses can easily scratch coatings and even the glass, so it's important to keep your binoculars dust free. When not in use, and that includes when you're snacking or eating lunch with the binoculars around your neck, keep the optical lenses covered with the lens caps or rain guard.

I try always to keep a photographer's lens brush in my birding jacket so when I notice dust on my lenses I can blow and brush it off quickly without the risk of grinding it into the lens with a tissue or cloth. I clean my binoculars at most once or twice a month when I'm birding intensively. When I clean them, I always first use the lens brush to gently sweep and blow particles away. If the eyecups are removable, you can unscrew them to brush everything completely off the lens.

After removing all dust, I moisten the corner of a high-quality lens cloth with a lens cleaning solution made specifically for coated binocular lenses (never use window cleaner!) and softly wipe the lenses clean. Then I buff the lenses with the dry portion of the cloth.

WHAT TO DO WITH OLD BINOCULARS

🐦 **Many birders keep their old optics** on a closet shelf just in case anything happens to their new ones. As insurance, this isn't a bad idea, but if your old optics are in usable condition, you may want to put them to work so that others can enjoy birds and protect their future. How? Donate your old optics to a local nature center or birding club, or to an organization such as the American Birding Association's Birder's Exchange, or to Optics for the Tropics.

Both organizations send used (and sometimes new!) binoculars to researchers and educators in Latin America and the Caribbean. Information about Birder's Exchange is on the ABA website at *www.americanbirding.org/bex,* and about Optics for the Tropics at *www.opticsforthetropics.org*.

Beyond Binoculars: Do I Need Fancy Equipment?

Q **What makes a spotting scope different from a telescope? And do I really need one?**

A spotting scope is a portable telescope that has been designed specifically for observing objects here on earth. The magnification of a spotting scope is typically on the order of 20× to 60×.

No one *needs* a spotting scope. But a spotting scope will make views of waterbirds, shorebirds, and grassland birds ever so much easier. It can even give you amazing close-up views of woodland birds, especially on or near their nests when you don't have to follow a moving object.

Q How do I choose a spotting scope?

A There are two basic scope designs: straight and angled. I used a straight scope for the first 30 years I birded. A straight scope makes it easy to home in on your bird unless it is well below you (as when you're birding on a ridge or bluff) or quite high up. Straight scopes are not easy to share with large groups, or even with family members, unless everyone is close to the same height. Tall people have to scrunch down to see through a scope set up for a short person.

I bought an angled scope in 2005. There was definitely a learning curve in figuring out how to get the bird into the view because you have to look down into the eyepiece to see a bird that is straight ahead. But now that I'm used to it, I'd never consider going back. Angled scopes, like porroprism binoculars, have one fewer glass element, and that gives them a slight edge over straight scopes optically. Also, when a bird is very high or very low, it's easy to rotate most angled scope models in their housing so you can more comfortably look through the eyepiece sideways. And when the scope is set up for the shortest person, the tallest ones don't have to scrunch down — they can simply bend over a bit to look down through the eyepiece.

Most birders use zoom eyepieces on spotting scopes. These are usually configured to magnify 20× to 60×. I usually prefer using a fixed 30× eyepiece, but the zooms are quite satisfactory for most people and have a broader value in a lot of birding situations.

The objective lens of most spotting scopes ranges in size from about 60 mm to about 85 mm. As is true of binoculars, spotting scopes with larger objective lenses gather more light than scopes with smaller ones. The less expensive the scope, the more important it is to get a larger lens to ensure your image is bright and crisp. If you can't afford the best, the rule is just like the rule for inexpensive binoculars — the bigger the objective lens and the lower the power, the clearer and brighter your view will be.

˅ ˅

Q How do field trip leaders get so fast at finding distant birds in their scopes?

A The more you scope birds, the faster you'll be. If you use a zoom eyepiece, always start out at the lowest magnification while you scan; once you've found the bird you can zoom in.

As with learning to use binoculars, practice on distant trees and other objects first. When I bought my first spotting scope, I started a list of all the birds I saw through it. It was slow going the first couple of weeks, but with the motivation of a growing list, soon I could find things in it amazingly fast!

Q When I go to my local birding club's website, I see zillions of photos taken by birders. I thought photographers needed all kinds of special equipment and to hide in blinds, but some of these pictures were taken by people on field trips. How can people take decent photos while keeping up with a field trip group?

A Many camera companies sell "extended zoom" cameras that can zoom in 15× or even more, with image stabilization so the photos are reasonably sharp. I take one of these cameras everywhere.

Another strategy for taking photos while birding is to hold a small digital camera close to a spotting scope. To hold the camera steady and at the optimal distance from the optical lens, some optics companies sell special adaptors, and some birders create makeshift adaptors from such things as cold-medicine measuring cups or the upper part of plastic vitamin bottles. I've taken some very high-quality photos this way. Taking digital pictures through a spotting scope in this way is called *digiscoping*. You can learn about techniques and see examples at *www.allaboutbirds.org*.

⌄ ⌄

Q I see lots of birders using electronic gadgets. What are they doing?

A Birders trying to keep up with rare bird sightings in their area can get updates by means of Internet Listservs, catching up on the latest posts on their iPhones, or through other devices. Some birding networks send out cell phone text messages when a rare bird is sighted. A lot of birders nowadays carry iPods, iPhones, or mp3 players with earbuds or a small speaker. Sometimes they review bird songs while trying to identify birds, and sometimes they play recordings in the field to try to lure birds in for a closer look; this is called *playback*. I've done this a few times, and it can be amazingly effective, but it can also be disruptive for birds exhausted from migrating or busy with nesting responsibilities. Playback should never be done to call up hotline rarities that dozens or hundreds of other birders are trying to find — this kind of constant disruption is suspected of causing nest failures, making birds fly off to quieter places, and even leading to a bird's death. It also should never be done in popular birding locations where there are plenty of disruptions anyway.

Other electronic gadgets that birders might carry in the field include recording equipment — parabolic or shotgun microphones and recorders. Now that it's possible to get reasonably priced digital recorders not much bigger than a deck of playing cards, more and more birders may start making their own high-quality sound recordings. Of course, many birders also carry cameras that take photographs and/or video.

Learning More about Birds

Q **Is there any equipment a birder should have besides binoculars, a field guide, and maybe a spotting scope?**

A Yes. Every birder should always carry a field notebook and pen or pencil. A field notebook is indispensible for keeping track of what you see — not just for listing each species but also noting how many you see and any interesting behaviors you observe, the places you go, and weather conditions. When you're starting out, you may want to record what field marks you note on each bird, especially unfamiliar species. It takes a bit of discipline to get into the habit of writing things down every 15 minutes or so when on an outing, and sometimes when you go out wanting to feel as free as a bird, a notebook may make you feel encumbered. But you'll soon be grateful for the memories and will have good habits when you encounter a rarity that requires documentation.

Birding stores sell waterproof spiral notebooks, but less expensive small notebooks can work well under most field conditions. To minimize weight, some people keep a small ring binder at home and tuck just a few sheets, folded if necessary, into their field guide. A daily checklist sheet can be useful if it leaves plenty of room for keeping track of numbers for each species and for notes. Your notes will be fun for you to review in coming years. You can multiply their value many times over if you also enter the data at *www.ebird.org*.

⌄ ⌄

Q There are so many field guides to choose from! How do I pick one?

A Begin by browsing the field guides at the library or bookstore to get a sense of which one works the best for you. Most experienced birders prefer a field guide with drawings by an expert rather than one with photographs. Good bird artists portray birds in similar poses, using their experience and knowledge to make it easier for you to key in on the important field marks. With photographs, lighting conditions and differences in bird postures can obscure important features or highlight unimportant ones, although the photos in some well-done guides are digitally manipulated to make color comparisons among different species more accurate.

Size is very important with a field guide, because if your book is too large, you won't want to carry it in the field, but if it's too small, it may not include all the birds you're likely to see in your area. If you hope to eventually become proficient at birding, it's wise to start with a guide that shows all the birds of North America or at least all the birds of the East or the West.

Hawaii's Birds, a small guide published by Hawaii Audubon, is the only field guide with complete coverage for a single state. Other than that one, I never recommend using guides that show the birds of a single small area — almost every beginner sees at least a few species in the first several months of birding that aren't included in more minimalist guides, leading to misidentifications and frustration.

Keep these things in mind as you browse through several field guides, and pick a few that seem best on an overview. Now look up two or three birds that you're very familiar with in each one. In your judgment, which seem closest to how you've experienced those birds? Consider color and poses. Also, how easy is it to find each of these familiar birds in the book? Remember: With field guides as with optics, there is no "best." Beyond a few basic issues, it's a matter of personal preference.

v v v v v v v v v v v v v v v v v v v v v v v v v

Q **I can find birds I know in a field guide by using the index. But if I don't know what the name is to use the index, I can never seem to find the bird in the book before it flies off. What am I doing wrong?**

A The birds in most field guides are not organized alphabetically or by color, but according to how they are related to one another. Probably the single most important thing to do when you buy a field guide is to read it cover to cover, familiarizing yourself with each group of birds. Pick your book up several times a day and thumb through it, paying attention to the shapes and relative sizes of the birds, color patterns, and notes about behaviors and habitat.

You may notice that many swimming birds that seem to be shaped like ducks aren't grouped with the ducks, geese, and swans. Grebes, loons, coots, pelicans, cormorants, and even tiny shorebirds called phalaropes all swim in a ducklike manner. But despite the superficial similarities that all these birds share, they represent several different genetic lines so are found in different sections of most field guides.

Although color patterns are important, being able to place birds in families is even more critical. Many birds have all-white plumage except for black wingtips, including American White Pelicans, Snow Geese, Whooping Cranes, White Ibises, and gulls; if you see one of these, its color pattern won't be very useful until you get to the pages that show its family. And in poor light, you may not get a good sense of color at all. So shape and posture are usually the first features to pay attention to.

As you flip through your field guide and come across an interesting species, don't read just the species account — go back and read that bird's family profile as well, and notice how it compares and contrasts with its close relatives. By doing this frequently, your guide will soon become a familiar friend, and you'll be able to recognize bird families, which is an important step toward mastering bird identification. The more you do this at home, the quicker you'll be at identifying birds in the field.

˅ ˅

Q Why don't they organize field guides by color? That would make things so much easier!

A If field guides were organized by color, it would certainly make it easy to find birds that are only one color, such as swans or cardinals or crows! But where would the authors place a bluebird? Many times, of course, we see the brilliant blue back, but from the front, a perched Eastern or Western bluebird appears far more red than blue. Is a Great Black-backed Gull black or white? Would a Red-winged Blackbird go in the black section or in the red section? Is a Painted Bunting red, blue, or green? If we were to choose all the conspicuous colors for each bird and show them in each of the relevant color sections, the book would be too heavy to call a field guide!

∨ ∨

Q **While cleaning my grandfather's attic, I found several of his notebooks, filled with bird lists and details about where and when he saw the birds. Some of the records went back to the 1940s! Would this information be useful to anybody?**

A Absolutely. Ornithologists are keenly interested in long-term data about birds, especially if it includes information about the number of birds seen. However, the data can be useful only if they're in a format that researchers can access and then analyze the information. I hope you'll consider entering your grandfather's data into eBird at *www.ebird.org.* A project of the Cornell Lab of Ornithology and the National Audubon Society, eBird enables birders to contribute their sightings to a permanent database online. The information then becomes accessible to scientists and birders alike through maps, graphs, and tables.

Birders who have kept electronic records in other formats can upload their sightings to eBird as well. Anyone can enter sightings of birds at any time, whether from this morning's bird-watching trip or from a list from decades ago.

∧∨

The Ins and Outs of Bird-Watching

Q **I have so many problems spotting birds in the woods. I can hear plenty of them in the trees but usually can't see even one! What am I doing wrong?**

A The thick branches and foliage of trees and shrubs makes forest birding a challenge. You'll need to be patient to be successful. Get into the habit of looking along every branch, and if you're hearing any voices, looking patiently through the branches for the singer. Sometimes it's hard to gauge direction and distance from a bird by sound unless we walk around a bit — making a couple of steps to your right and left may help you figure out more precisely where the bird is.

Some birds flit about as they sing, so the sound will be moving with the bird. Some birds hold still for many minutes while singing. If you can't pick out their shape, you're not going to see these until they fly off. If these birds are at treetop level, they can be especially tricky to find because so many leaves and branches obscure them. Looking around for better vantage points — a big boulder or tree stump to stand on, a little opening giving you a wider view — will help.

During migration, warblers, vireos, kinglets, and other migrants often seek out chickadees to associate with. Chickadees know every little thing about their woods, and don't mind company, so migrants associating with them can more effectively find food and elude predators. It's a good idea to listen for chickadees, paying special attention to any little birds moving about when chickadees are around.

Lower-power binoculars are especially useful in forests. They give a wider field of view than higher-power ones, so there's at least a little higher probability that "your" bird will be in view; and for the same size, they provide a brighter image than higher-power glasses, which is especially useful in shady areas with a dense canopy.

Birders in forests often try "pishing," which means making *spish spish spish* sounds that may sound like a baby bird in distress. Pishing sometimes brings birds in for a momentary view. You still have to be alert, because when birds come in to check you out, they may still be concealed in foliage.

Unfairly, at least from the standpoint of a novice, the more birds you have seen in forests, or any other habitat, the more birds you will be able to see. The first ones can be ever so tricky, but as you get to know each one, you'll recognize the kinds of vegetation and where within it each species is most often found. With every new bird you find, finding the next one will be that much easier.

Don't get discouraged when you go out with an experienced birder who calls out all kinds of birds you don't see. Most experienced birders recognize bird voices, and most of the birds they call out were heard but not seen. And don't be afraid to ask them for advice and to point the birds out. Almost all will be happy to share their expertise.

Q Is it better to go birding on my own or with a group?

A Attending local birding field trips or going birding with more experienced birders can provide a lot of shortcuts for beginners trying to become proficient. You'll learn the best hot spots, glean all kinds of valuable tips from experts, test out spotting scopes, and build up your life list much more quickly than you could on your own. And the fun of birding can be enhanced by the camaraderie of like-minded enthusiasts.

That said, in my opinion you should spend at least as much time birding alone as you do on organized field trips. On field trips, novice birders have a tendency to defer to experts and those with more experience, and even to shut off their own sense of inquiry when the answers are so readily available. The more time you spend teasing out frustrating identifications on your own, the more skilled you'll become. Another advantage of going solo is that you can explore places that aren't as popular, and occasionally discover "good birds" that no one else would have found.

A birding buddy can enhance your birding experiences wonderfully. The trick is finding the right one. The ideal birding companion shares your birding rhythms and your financial and time constraints, shares your level of competitiveness, is roughly your equal in skills so you can learn from one another and egg each other on to higher skill levels, and is fun to be with on long trips. Ideal birding buddies are hard to find! But if you can find a group of like-minded birders who enjoy being together from time to time, you can enjoy convivial company and save money and gas when headed to fairly distant birding sites or chasing hotline birds.

ˇ ˇ

Q Some people say we shouldn't worry about what colors we wear when birding. Others say not to wear bright colors, and some say we shouldn't wear white but anything else is okay. I noticed when people were searching for the Ivory-billed Woodpecker, they were all wearing camouflage. Who should I believe?

A I've seen plenty of birds when wearing bright colors, including hummingbirds that were attracted by my bright red hat, an oriole that was drawn in by my bright orange University of Illinois sweatshirt, and a Red-winged Blackbird that waged a brief but fierce territorial battle with my red-striped bicycle helmet. That affirms that birds do, indeed, notice bright colors. But keep in mind that for every bird attracted to a bright color, another bird might have been alarmed after spotting me and stayed out of sight.

Many birders consider white the single worst color to wear because pure white is so conspicuous. Many animals, from white-tailed deer and cottontail rabbits to Dark-eyed Juncos, take advantage of the conspicuousness of white to flash it when running away, which may serve as a warning to others in its group to also flee. Remember, it's not the color that birds are afraid of — it's you. So it's probably prudent to wear inconspicuous field clothing when possible.

Of course, even if we're wearing camouflage, it's hard to believe that birds, with their superior senses of vision and hearing, aren't aware of our presence; although most bird photographers affirm that they get the most reliable opportunities for photos when they blend into the background as much as possible. Many field trip leaders ask that participants dress in dull colors. Whether or not you agree with their reasoning, it's a courtesy to comply.

Much more important than clothing choices are walking fairly slowly, not making sudden quick movements, keeping your voice down, and stopping often to quietly look and listen.

⌄ ⌄

Q I saw a rare bird but no one in the local birding club believes me. How come birders are so arrogant?

A Don't feel bad — birders question even the most experienced birders among us! It isn't a matter of arrogance; it's a matter of ensuring that every single bird recorded by a state or local birding organization has been identified with as close to 100 percent certainty as humanly possible. To get a rare bird included in the body of records for most birding or ornithological organizations, you need to clearly show that you've seen all the critical field marks and excluded every possible similar bird. Because it's becoming so easy to take photographs in the field, more and more organizations are now requiring photographic evidence to accept exceptionally rare or new species on their lists.

THE ALL-IMPORTANT LIFE LIST

A life list is the list of all the wild bird species seen during a birder's life. Most birders count only birds they see in nature, not zoo or pet birds, or birds at wildlife rehabilitation facilities. The competitive sport of birding involves seeking out and identifying birds in the natural world. Birds that are captive don't meet that requirement.

Of course, your bird lists are your own personal business, and if you want to count birds you've seen in zoos and aviaries, or even birds you've seen on TV, you can do so. But when some-one asks, "What's your life list at?" or "How many birds have you seen?" your answer won't necessarily make sense to anyone but yourself if you aren't following the rules of the American Birding Association (ABA).

The ABA keeps track of birding lists for people who want to bird competitively. At the top of their 2007 list is Tom Gullick of Spain, who has seen 8,702 birds in the world. Macklin Smith of Michigan tops the North American (north of the Mexican border) list with 876 species, and also tops the list of birds seen in the United States including Hawaii with 921 species. (Canada has very few birds not found in the continental United States and Alaska, but Hawaii adds a great many to the list — not only native species but also the many birds that have been introduced and become established there, as well the pelagic birds seen along the coast.)

The ABA's annual lists of records, including top lists for every U.S. state and Canadian province and every continent and coun-try, can be found on the Internet, indexed at *www.aba.org/bigday*. To make a level playing field, birders have to follow the same rules of listing, which are available on the website.

Q What should I do when I see a rare bird?

A While you're in the field, write down every single field mark, posture, and behavior you observe, as well as size comparisons with objects or other birds near the rarity. It's best if you can draw or, better yet, photograph the bird. Write down your thought process: how did you come to think it was X rather than Y or Z, and be sure to explain how you excluded every similar species. If your state has a rare bird report form, keep a copy of it tucked in your field guide for just this occasion and fill it out as you watch the bird.

Don't shy away from using your field guide after getting a thorough look at the bird, but be sure to read the text to make sure you're noticing all the important field marks and comparing your bird with every similar species. Advanced birders say you'll lose credibility if you say that your bird "looked exactly like the picture in the book." If you do compare your bird to a field guide illustration, be sure to point out the specific ways that your bird did look like the picture, but also how it differed.

After the bird leaves, check your references to ensure that you haven't missed anything you should have seen. When you are certain about your identification, let your local birding hotline know about it unless the bird may be vulnerable to spooking under pressure by a lot of birders gawking at it. Also, report the bird to *www.ebird.org* and your local and state birding organizations. Depending on how rare the sighting is, you may need all those notes you took!

⌄ ⌄

Q I've read about places in the United States where more than a hundred thousand hawks fly past in a single day, and about a place in Mexico where more than a million birds may fly past in a single day. How is it possible to count so many?

A I can give you an example of a count I witnessed on September 15, 2003, at Hawk Ridge Bird Observatory in Duluth, Minnesota, when a total of 102,329 hawks were counted. Fully 101,698 of them were Broad-winged Hawks. There were also 445 Sharp-shinned Hawks and 83 American Kestrels. No more than 35 individuals were counted of any other species. How did they count them all and keep the species straight?

That day there were several counters, each in charge of one area of the sky, and each counter had a volunteer assistant recording birds. Hawk Ridge is a bluff a mile from Lake Superior. One counter was in charge of counting birds flying above the lake, with sky and water in the background. Another counted birds below the ridge, with houses and trees in the background. One took the birds flying overhead, and one took the birds along the far side of the ridge, opposite the lake. The birds weren't crossing paths because they were all moving parallel to the lake, but other experienced birders at the count station were keeping track of the overall movement patterns to alert the counters if there were any shifting paths so birds wouldn't be counted twice.

The counters counted Broad-winged Hawks, sometimes by tens and sometimes even by hundreds, and then called out the final number for each *kettle* (a swirling mass of hawks numbering anywhere from about five to many thousands) to the assistant who wrote down the number. Meanwhile, as Sharp-shinned Hawks passed by, the counters clicked them on a mechanical clicker. Every hour that number was recorded and the clicker reset to zero. As the counter was counting these two species, if a bird of another species went by, the counter called it out for the assistant to record individually.

That was, as of the end of the 2008 hawk-counting season, the biggest day ever at Hawk Ridge. The second largest flight was recorded on September 18, 1993, when 49,548 were counted. Hawk Ridge averages about 100,000 or so hawks through an entire season from August until November. The Hazel Bazemore Hawk Watch in Corpus Christi, Texas, pretty much directly south of there, often reports single day totals of more than 100,000 hawks — sometimes as many as 400,000 — and averages about 720,000 hawks a year. The record flight there happened on September 28, 2004, with a total day's count of 520,267 raptors including 13 species. Of that day's count, 519,948 were Broad-winged Hawks.

❧ ❧

Q I want to plan a trip to see migrants along the Gulf Coast in Texas next spring. How can I schedule this so I'm there on the best days?

A April is an ideal month along the Texas coast. Migrant songbirds begin arriving in large numbers early in the

> Nearly every Broad-winged and Swainson's Hawk in the world passes through Veracruz, Mexico, in a 15-day period each fall. Single day counts can surpass a million! And an even larger migration passes over the Panama Canal, where all the hawks flying from North to South America must funnel through.

month, but maximum numbers of warblers fly in usually around the third week of April. On days with "weather events," often after southerly winds in Mexico sent birds flying across the Gulf of Mexico but rain on the Texas coast stops them, numbers can be huge, with as many as 30 species of warblers!

Unless you happen to have your own private jet and can take off at the drop of a hat when favorable weather is about to arrive, there's a huge amount of luck involved in being at any migratory hot spot on the best days of the season. Part of your strategy should be to find alternative birding plans if one spot doesn't pan out. Learning about nearby birding spots ahead of time can provide an excellent "Plan B." The American Birding Association publishes birders' guides to many hot spots, which make researching the possibilities for any birding adventure fun and easy. Remember that migrants crossing the Gulf from Mexico often don't arrive on the upper Texas coast until afternoon. So even if few birds are around in the morning, it's often worth checking "migrant traps" again in the afternoon.

˅ ˅

COUNTING MIGRATING FLOCKS

 Counting large numbers of birds is yet another birding skill for which practice makes perfect. It's easiest to start out counting smaller groups of birds that are perched or on the ground, and work your way up. Small flocks of a single species can be counted one by one. Larger flocks often must be counted in groups. Count by the smallest grouping you can.

In most places you'll usually be able to count single species flocks by fives or tens, but when you are at a major gathering place on a shore or wetland, you may have to learn to count by 20s or even 100s. If it's a huge gathering, divide up the total flock into sections, count one section, and multiply by the number of sections to get a reasonable estimate.

To count flying birds, again practice on smaller groups. Block off a group of individuals, count them, and then extrapolate to the entire flock; or count birds per unit of time as they pass a specific point.

Q My daughter loves birds. What kinds of careers might be good for her?

A There are many branches of science she may want to explore, including ornithology, wildlife biology, ecology, conservation biology, behavior sciences, avian physiology, or veterinary medicine.

She might consider elementary education — in the proper hands, bird study can be integrated wonderfully into a multi-disciplinary program, and curriculum aids such as the Cornell Lab of Ornithology's BirdSleuth program and Journey North's online lessons can make teaching about birds much easier. High school biology teachers can introduce their students to birds and inspire them to learn about science.

If she has a legal mind, she may want to be an environmental lawyer with a background in bird conservation. A love of writing and of birds can be a great combination for a career as an environmental author, journalist, or radio host. If she wants to work outdoors, she may be able to do fieldwork for government agencies or nonprofits, or lead birding tours.

Blue Jays lower their crests when they are feeding peacefully with family and flock members or while tending to nestlings.

KEEPING AN EYE ON CROWS

In some areas, the American Crow leads a double life. It maintains a territory year-round in which the entire extended family lives and forages together. But during much of the year, individual crows leave the home territory to join large flocks at dumps and agricultural fields, and to sleep in large roosts in winter. Family members go together to the flocks but do not stay together in the crowd. A crow may spend part of the day at home with its family in town and the rest with a flock feeding on waste grain out in the country.

Communal crow roosts can range from a few hundred to two million crows! Some roosts have been forming in the same general area for well over 100 years. In the last few decades some of these roosts have moved into urban areas where the noise and mess cause conflicts with people.

Sometimes they conflict with people's pets as well. I used to watch three crows that would arrive in my neighbor's yard the moment she left for work every morning. As she headed out to the car, she would tie her Springer Spaniel up and give the fairly large dog a bowl of food. No sooner did she back her car out than the crows would drop down and walk slowly toward the dog, staring intently at her. The dog meekly backed away from her dish, and the crows stuffed their throat pouches with dog food before leaving for the day.

AMERICAN BIRDING ASSOCIATION'S PRINCIPLES OF BIRDING ETHICS

Birds often go about their business as usual even as you watch them, but birding, especially by large groups or in sensitive or popular areas, can cause stress for birds and can sometimes even damage their habitat. Birders should be especially sensitive around nests, since distressed birds may draw the attention of nearby predators.

The American Birding Association (*www.aba.org*) has created a set of guidelines to address situations in which birds may be harmed. Their Code of Birding Ethics emphasizes that "In any conflict of interest between birds and birders, the welfare of the birds and their environment comes first."

Code of Birding Ethics

1. Promote the welfare of birds and their environment.

1(a). Support the protection of important bird habitat.

1(b). To avoid stressing birds or exposing them to danger, exercise restraint and caution during observation, photography, sound recording, or filming.

Limit the use of recordings and other methods of attracting birds, and never use such methods in heavily birded areas, or for attracting any species that is Threatened, Endangered, or of Special Concern, or is rare in your local area.

Keep well back from nests and nesting colonies, roosts, display areas, and important feeding sites. In such sensitive areas, if there is a need for extended observation, photography, filming, or recording, try to use a blind or hide, and take advantage of natural cover.

(continued)

Use artificial light sparingly for filming or photography, especially for close-ups.

1(c). Before advertising the presence of a rare bird, evaluate the potential for disturbance to the bird, its surroundings, and other people in the area, and proceed only if access can be controlled, disturbance minimized, and permission has been obtained from private land-owners. The sites of rare nesting birds should be divulged only to the proper conservation authorities.

1(d). Stay on roads, trails, and paths where they exist; otherwise keep habitat disturbance to a minimum.

2. Respect the law, and the rights of others.

2(a). Do not enter private property without the owner's explicit permission.

2(b). Follow all laws, rules, and regulations governing use of roads and public areas, both at home and abroad.

2(c). Practice common courtesy in contacts with other people. Your exemplary behavior will generate goodwill with birders and non-birders alike.

3. Ensure that feeders, nest structures, and other artificial bird environments are safe.

3(a). Keep dispensers, water, and food clean, and free of decay or disease. It is important to feed birds continually during harsh weather.

3(b). Maintain and clean nest structures regularly.

3(c). If you are attracting birds to an area, ensure the birds are not exposed to predation from cats and other domestic animals, or dangers posed by artificial hazards.

4. Group birding, whether organized or impromptu, requires special care.

Each individual in the group, in addition to the obligations spelled out in Items #1 and #2, has responsibilities as a Group Member.

4(a). Respect the interests, rights, and skills of fellow birders, as well as people participating in other legitimate outdoor activities. Freely share your knowledge and experience, except where code 1(c) applies. Be especially helpful to beginning birders.

4(b). If you witness unethical birding behavior, assess the situation, and intervene if you think it prudent. When interceding, inform the person(s) of the inappropriate action, and attempt, within reason, to have it stopped. If the behavior continues, document it, and notify appropriate individuals or organizations.

Group Leader Responsibilities [amateur and professional trips and tours]

4(c). Be an exemplary ethical role model for the group. Teach through word and example.

4(d). Keep groups to a size that limits impact on the environment, and does not interfere with others using the same area.

4(e). Ensure everyone in the group knows of and practices this code.

4(f). Learn and inform the group of any special circumstances applicable to the areas being visited (e.g., no tape recorders allowed).

4(g). Acknowledge that professional tour companies bear a special responsibility to place the welfare of birds and the benefits of public knowledge ahead of the company's commercial interests. Ideally, leaders should keep track of tour sightings, document unusual occurrences, and submit records to appropriate organizations.
Please follow this code and distribute and teach it to others.

CHAPTER 3

Living with Our Feathered Friends: Solving Bird Problems

As wonderful as birds are, when the natural world collides with our own, sometimes it's tricky figuring out how to solve the problems: Cardinals shadowboxing with their reflection on our windows. Woodpeckers carving up wood siding. Geese crowding onto the golf green. If I had a nickel for every time someone called my house asking me how to solve a bird-related problem, I'd be wealthy! Some issues are harder than others, but most of the time we can solve bird problems without harming birds. How? I'm glad you asked!

᷍᷍

The Birds — Not Just a Movie

Q For the past few weeks, I've woken up to the sound of a woodpecker hammering on the house. I can see where it has been drilling holes under the eaves. I've tried chasing it away but it keeps coming back. Why does it keep eating my house?

A Woodpeckers don't actually eat houses, though that is hardly a comforting technicality when they're carving holes into your wood siding. They *do* eat insects in the wood, and they also chop holes into houses when trying to excavate cavities where they can nest, roost, or store their food. If you have the right type of house, they may also drum on it to defend their territories or to attract mates.

Woodpeckers declare their territory by hammering a specific rhythm on the most resonant structure they can find, often beginning early in the morning. Drumming can be loud and annoying, but it doesn't cause serious damage — usually small dents in the wood, grouped in clusters along the corners or on fascia and trim boards. The holes may sometimes be as large as an inch across, round, cone-shaped, and generally shallow.

If the hammering isn't particularly loud or rhythmic, the woodpeckers may be looking for food. Sometimes insects work their way into siding, especially grooved plywood siding. Leaf cutter and carpenter bees, grass bagworms, and other insects crawl into grooves in the siding; when woodpeckers hear them, they cut into the wood to get them. Holes made when hunting for insects tend to be small and clustered, usually three to six or so, often in a line.

Woodpeckers also sometimes dig holes in houses to make a place to roost or nest, usually on dark-stained or natural wood houses near wooded areas. These holes are usually dug in springtime and are large enough for the woodpecker to fit into.

Acorn Woodpeckers store food in holes. These holes are often about the size you'd expect and, sure enough, often contain an incriminating acorn. Fortunately for all of us, they are far more likely to do this to trees than houses.

WOODPECKER DAMAGE STUDY

A 2007 Cornell Lab of Ornithology study surveyed 1,400 homes in the area around Ithaca, New York, to learn which ones were most enticing to woodpeckers. It also tested six common long-term deterrents to see how effectively each prevented wood-pecker damage.

The study affirmed that damage was most likely on homes infested with ants or carpenter bees. It also found that homes with vinyl or aluminum siding, or painted in light colors, were less likely to be attacked than houses with stained wood siding or siding painted a dark color.

If your house attracts woodpeckers, what do you do about it? The study tested deterrents intended to scare away the wood-peckers, such as life-sized plastic owls with paper wings, reflective streamers, plastic eyes strung on fishing line, and a broadcast system that played woodpecker distress calls and hawk calls. The study also provided alternatives for the woodpeckers, such as ready-made roost boxes and suet feeders. The only deterrent that worked consistently was the plastic streamers, but even they stopped the damage at only 50 percent of the houses.

At least half of the time, addressing woodpecker issues involves more than just trying to scare them away. The Cornell Lab has a great resource to help you identify what the wood-peckers are doing and then find an effective solution. Look for Woodpeckers: Damage, Prevention and Control at *www.birds.cornell.edu/wp_about*. Suggestions include using wood putty to fill nest and roost holes as soon as they appear and then discouraging woodpeckers from reappearing by covering the area with burlap for a few weeks. Burlap will also discourage drumming, since it dulls the sound.

The problem with insect-infested wood is more difficult, because woodpeckers will continue to come as long as your house harbors food for them. So you have to first get rid of the insects, then completely repair the holes with wood putty, in order to solve that problem.

Q For the last three winters, Blue Jays have been pecking at the south-facing front of our house, removing most of the paint by the end of the winter. Why do they do this, and is there any way we can get them to stop? They're not only destructive, but it's very annoying to have them hammering on the house every morning at daybreak!

A During the winter of 2000–2001, Deborah Jasak, a Project FeederWatch participant, called the Cornell Lab of Ornithology to report the same problem with Blue Jays taking paint chips from her New Hampshire house. When FeederWatch staff asked others if they'd had the same problem, they were surprised to hear just how widespread it is. After the *Boston Globe* ran a story about the issue, Massachusetts Audubon received 160 reports of Blue Jays chipping and eating paint from houses.

Why do they do it? Paint manufacturers use calcium carbonate, or limestone, as an extender pigment in paint, making paint a source of calcium. Another FeederWatch research project found that Blue Jays consumed more than twice as much calcium as other birds do. And they seem to take paint from houses mostly in the Northeast, where soils are unusually low in calcium. When Deborah Jasak put out eggshells, the jays started taking those instead and left her house alone. She tried other sources of calcium, such as oyster shells, sand, dirt, and trace minerals, but eggshells were the only effective deterrent, and if they got buried under snow, the jays returned to peeling paint off the house.

In rare cases, jays may take the eggshells yet continue to peel house paint. These intelligent, social birds develop habits that can be hard to break, and may simply enjoy the activity of paint peeling. You might try covering the painted area they're damaging with screening or burlap, or hanging shiny helium balloons to float in that area. When I rehabbed birds, I learned that most birds, especially Blue Jays, seemed terrified of helium balloons and their unpredictable movements.

SUSPICIOUS THIEVES

Some Western Scrub-Jays search out their food items on their own, while others raid the food caches stored by other scrub-jays and also by Acorn Woodpeckers and Clark's Nutcrackers. Researchers have found that when "thieving" scrub-jays hide their food items, they spend a lot of time looking around to check if other jays are watching them; nonthieving scrub-jays seem unsuspicious and don't look around before hiding their own food. The more a scrub-jay engages in stealing food from others, the more suspicious it becomes.

Some animals seem to love it when scrub-jays raid them for food — if that food happens to be parasites. Western Scrub-Jays frequently stand on the backs of mule deer picking off and eating ticks. The deer seem to appreciate the help, often standing still and holding up their ears to give the jays access. If you've read Jack London stories, you may have noticed that he mentions "moose birds." These are Gray Jays, which acquired that nickname because they sometimes pick the parasites off moose.

Q A cardinal keeps flying into my kitchen window. I think he's doing it on purpose, and it's driving me nuts. How can I make him stop?

A Many territorial birds are incensed when they discover another bird of their species and sex on their territory. Reflections in windows and auto mirrors can appear to be exactly this. Cardinals and robins are the species most likely to start attacking their window reflections, and the attackers are usually males. Sometimes female cardinals or robins will do this, and other species occasionally attack their reflections, too.

In nature, when a cardinal discovers another cardinal on his territory, he may first respond by making a warning call or fly to a perch and sing, or he may instantly lower his crest and make *pee-too* or *chuck* call notes. If the intruder doesn't leave, he'll lower his body, open his mouth, vibrate his wings, and make various other calls. If the other cardinal still doesn't leave, he lunges, but usually the intruder escapes before it comes to blows. Wild cardinals often countersing with their neigh-bors, which may give an outlet for some territorial disputes without the birds resorting to physical battles. Wild cardinals have engaged in fighting and continual chases for as long as 30 minutes, but most physically aggressive encounters last only a few seconds.

Reflections don't leave, however, and they don't counter-sing. To drive the reflected bird away, the cardinal gets into full battle mode, lunging at the glass. But rather than flying away or engaging in a normal "chase," the reflected bird matches every aggressive posture, and when the real cardinal hits, the window is unyielding. The real cardinal gets so intent on driving the reflection away that he may waste weeks fighting with it.

It's not too hard to solve the problem if the cardinal is fixated on just one window. Just soap the outside of the window or cover it with screening or newspapers for a few days. Unfortunately, sometimes when a reflection disappears from one window, the cardinal searches for and finds it in other windows.

Sometimes you can scare a bird away by taping helium balloons or shiny streamers on the window. Their unpredictable movements may scare birds from approaching close enough to

MIRROR, MIRROR, ON THE WALL

Some birds, especially in the crow family, can probably recognize that a mirror reflection is just that, not a real bird. One experiment showed that magpies marked with bright yellow or red on the throat reacted to their mirror image by scratching at the colored area on their own body; birds marked with black matching their throat feathers didn't react that way. This ability to recognize oneself in a mirror isn't known to occur in other animals except primates, dolphins, and elephants. It's unfortunate that cardinals and robins don't share this ability!

see their reflection. Usually by the time the helium is gone and the balloons sink, the bird will have moved on. Some people recommend using Great Horned Owl decoys to scare birds away, but I once saw a bird using one of those as a perch from which to more conveniently continue to attack the window.

No matter what you do, the cardinal won't attack your windows forever; he'll usually lose interest when the breeding season advances and passions aren't running so high.

TURNING LEMONS INTO LEMONADE

Sometimes when people are faced with a bothersome situation, they turn it into a good thing. I had a friend who was dismayed when a Great Blue Heron started feeding in her pond, but she later decided that the beautiful heron enhanced her yard even more than the fish. Another friend invented complicated strategies for keeping squirrels out of his feeders, but none worked. He finally switched to devising feeders to challenge squirrels, rather than to exclude them. He got to know the individual squirrels in his yard, and now finds them as entertaining to watch as the birds.

When waxwings, robins, and other birds started devouring the fruit on my husband's beloved cherry trees, he quickly noticed that the birds concentrated in the top branches. He decided that it was much easier to pick the lower branches anyway. No need to pull out the ladder anymore, he has company while he picks, and we still freeze enough cherries to last until the next year.

HOW TO PREVENT BIRDS FROM COLLIDING WITH WINDOWS

Of all the troubling issues facing birds in our backyards, windows are one of the most devastating. Current estimates are that every year, worldwide, billions of birds are killed in collisions with window glass. Some crash into lighted windows on tall buildings at nighttime during migration, but a great many collide with windows on our own houses. What can we do to reduce the kill?

There are two different strategies for protecting birds from glass: to make the glass more visible to avoid collisions in the first place, and to make collisions less lethal by placing screening in front of glass.

To make glass more visible, you can try window coverings that are opaque from the outside but provide a good view from the inside. Decals placed on the outer glass are also effective, as long as they're placed very close together, separated by only 2–4 inches (5–10 cm). Streamers, sun-catchers, or other decorative objects only work if they, too, are placed close enough together and on the outside.

The problem with either of these is that if you set out enough of them to be effective, you can obstruct your own view. Some new decals appear almost clear to our eyes, but because they are visible in the ultraviolet range, birds can easily see them. Again, these must be closely spaced on the outside of the glass to be effective.

Lines drawn on windows with highlighter markers and "invisible markers" that glow under black lights are easily seen by birds and can stop bird collisions if drawn in a tight grid pattern with openings no wider than about 4 inches (10 cm) high and 2 inches (5 cm) wide. To take advantage of the ultraviolet properties of the markers, they need to be used on the outside of the glass, since glass filters out UV light. You must also redraw them frequently because UV inks fade noticeably in less than a week.

You can also make your windows more visible by setting your feeders on the glass or window frame. Birds feeding right there may notice the glass and avoid flying into it. If they do take off toward the glass, especially when an unexpected predator suddenly appears, their speed on takeoff is usually too slow to cause injury if they collide with the glass. It takes only a couple of wing beats before birds are going full speed, so feeders farther than about 6 feet (2 m) from windows are far more dangerous than those right by the window.

To make collisions less dangerous, netting or screening attached to the window frame must be taut and set a couple of inches in front of the glass, so the bird will bounce off before it reaches the glass. Window screen or garden netting can work. Make sure it's taut enough to serve as a trampoline.

You can learn more about preventing window collisions from the Cornell Lab of Ornithology at *www.birds.cornell.edu/ Publications/Birdscope/Summer2008/window_screening.html*.

Q A phoebe is trying to build a nest on my porch light. We use the light a lot at night, and last year when a phoebe nested there, the eggs never hatched. We suspect they got too hot because of the light, so we really want her to build somewhere else. What should we do?

A Your phoebe has decided, just as you have, that your neighborhood is the right place to raise children. And phoebes make a neighborhood a little better because they eat so many flying insects, including mosquitoes. But a nest on a front porch can certainly be inconvenient for us!

To discourage her from nesting on the light, you can wedge in an insulated work glove or something else that won't be damaged or cause a fire if it gets hot. To expedite her choosing an alternative site, consider building a nest shelf somewhere else on your house. Plans for building one designed by the Minnesota Department of Natural Resources are available at *www. learner.org/jnorth/images/graphics/n-r/robin_nestbox.gif*. This plan is designed to serve robins, phoebes, and Barn Swallows.

Monitoring a nest can be fascinating for children, and you can have all the insect-eating benefits of a backyard phoebe without the porch-nesting drawbacks.

❧ ❧

Q A mockingbird is singing right outside my window all night long. He doesn't stop! Short of changing the front yard landscape, what do I need to do to shut him up, at least until sunrise?

Mockingbirds who sing all night long tend to be young, still-unattached males or older males who have lost their mate, so the best way to quiet him is to entice a female mockingbird to your yard, too. He's already doing his best to accomplish this, though to the disappointment of both of you, he's not yet succeeded. The singing will end on its own, usually within a few days or weeks.

Strategies for dealing with the problem for the duration include shutting the sound out of your house, by either closing windows or using ear plugs, or sending him elsewhere. If you can pinpoint the tree he's singing from, you might place nylon window screen fabric or a fabric with a similar weave atop the tree to discourage him from perching there. Using bird netting risks entangling him and other birds.

The situation brings to mind Robert Frost's poem, "A Minor Bird":

> I have wished a bird would fly away,
> And not sing by my house all day;
> Have clapped my hands at him from the door
> When it seemed as if I could bear no more.
> The fault must partly have been in me.
> The bird was not to blame for his key.
> And of course there must be something wrong
> In wanting to silence any song.

You might want to substitute for the final two lines:

> But of course there must be something right
> About getting a decent sleep at night.

> The mockingbird was Thomas Jefferson's favorite bird. He wrote a lot about its amazing mimicry abilities and songs, and how England had nothing to compare with it, in his *Notes on the State of Virginia*. He also had a pet mockingbird named Dick who lived in the White House.

Q **If birds have eagle eyes, why do they crash into windows, power lines, and guy wires?**

A Window glass is not only clear; it's reflective. Sky and trees are mirrored in windows, and since there was no such thing as glass in the natural world for the millions of years that birds have been evolving, few wild birds have yet evolved any ability to notice it. Window glass may have been used in Italy nearly 3,000 years ago, but it wasn't common in England until the seventeenth century. Huge picture windows have become widespread only in recent decades, much too recently for birds to have developed mechanisms to avoid them. Conservative estimates put the number of birds killed by glass every year in the United States from 100 million to as many as one billion.

Unlike branches and other natural structures, power lines and guy wires are straight and relatively thin, so they apparently appear two-dimensional, making it difficult for birds to gauge their distance from one until they're crashing into it. There is little data in North America, but extrapolating from data in Europe, as many as 174 million birds may be killed in North

America each year by high-tension line collisions. Nocturnal migrants are attracted to the vicinity of communications towers and their guy wires by the lights necessary to warn planes of their presence. Again little data are available, but the numbers of birds killed at communications towers and their guy wires is estimated to be from 5 million to 50 million every year.

In places where power line kills are known to be high, little, inexpensive objects called flight diverters can be placed on the lines, significantly reducing the kill. Some of these are as simple as wire coiled into a loose cone shape; others are more complicated, swiveling or rotating in 3 to 5 mph winds. They're expensive to place on existing power lines but much less so to put on new lines as they're being installed. Because they are

BIRD ON A WIRE

Why is it birds can sit on electrical wires and not get zapped? To get zapped, they need to close the circuit so the current actually flows through their body. If you were standing on a metal ladder and touched a bird on a power line, you'd both get zapped, because the ladder touching the ground would close the circuit. A squirrel can safely run along a power line, but when it reaches the end, if it makes contact with the transformer while it's still touching the wire, it will be electrocuted. When large birds perch on or very close to transformers and power lines, they often get electrocuted. As a matter of fact, in some areas electrocution is the main cause of mortality for Harris's Hawks.

three-dimensional and in some cases make noticeable movements, birds detect them and gauge their distance from them much better than they gauge their distance from wires without them. Encourage your local power company to use them on new wires.

^v^v^v^v^v^v^v^v^v^v^v^v^v^v^v^v^v^v^v^v^v^v^v^v^v^v^v^v^v^v^v^v

Outdoor Hazards

Q The geese on our local golf course are driving us crazy. They hiss whenever we come near, and I've slipped on goose poop and fallen on my keister twice! It seems like there are way more of them than there used to be. What can we do about them?

A Some Canada Geese are strongly migratory; they breed in northern Canada and Alaska and are seldom an issue in urban areas. But one subspecies of Canada Goose that isn't highly migratory has indeed become a serious problem. Ironically, this subspecies, the "Giant" Canada Goose, was on the verge of extinction in the 1950s due to over-hunting and habitat loss. But in 1962 a small flock was discovered wintering in Rochester, Minnesota. Birds from this population were reintroduced to many towns and parks. At the same time, wildlife managers were introducing nonmigratory flocks of Canada Geese to many wildlife refuges throughout the northern tier of states. In most of these areas, no breeding population had existed before. Suddenly Canada Geese were flourishing.

Unfortunately, the ability of geese to digest grass, along with their preference for expansive lawns from which they can see predators approach, has drawn more and more of these human-acclimated "Giant" Canada Geese to urban areas, especially places like airports, golf courses, parks, campuses, and cemeteries. A manicured lawn by a river, stream, pond, or lake is an irresistible invitation to geese.

What's the best way to deal with them? Allowing native vegetation to grow along shorelines and minimizing turf grass will at least reduce habitat for them. Obviously this isn't possible at airports or golf courses. In those circumstances, one of the most effective methods of keeping geese at bay is to hire a herding dog and handler to regularly chase the geese off, especially in late winter and early spring to prevent nesting.

In smaller expanses such as individual small lawns, setting plastic netting atop the grass often works, though the netting can be expensive, and it has to be rolled up to mow; some people would rather deal with the geese.

Q **Help! A heron is eating the koi in my pond! How can I discourage the bird and keep my fish safe?**

A Great Blue Herons flock to koi or goldfish in ponds the way some people flock to sushi bars. They can't help themselves! And once they've discovered a great fishing spot, it's very tricky to get them to leave. Urban conservationist Rob Fergus writes, "Vigilance is required for homeowners who don't want their koi pond treated as a giant bird feeder by herons. Since herons are fairly territorial, if one shows up uninvited (but face it, a bright orange fish is a pretty good invitation!), you may be able to drive it away with a life-sized heron decoy, available at many yard or garden centers."

These decoys don't always work. Rob warns, "Don't leave the decoy out in the same place for too long, as herons will quickly learn that an unmoving bird isn't a threat. It's probably best to bring it out only when needed, and to move it to a new location at least once a day."

Don't bother with fake alli-gators floating in the water — they simply don't work as anything but islands for turtles to rest on. Alter-native solutions include setting netting above the pond or keeping a dog near the pond (assuming your dog doesn't develop a taste for fresh-caught koi).

Q My cat Garfield never hurts birds, but my neighbor keeps asking me to keep him in the house. Why are bird lovers so paranoid about cats?

A The exact number of birds killed by cats each year is unknown, but the most conservative authoritative estimates place the annual kill in the United States at close to a hundred million birds every year, and some careful studies place the kill at about half a billion birds every year. Feral cats are a huge problem, but housecats that spend part of their lives outdoors also kill significant numbers.

Cats are "natural killers"; their instincts lead them to hunt small, moving creatures. But they are not natural in the sense that they're not part of the animal life native to North America. They were brought here by humans and seldom survive to lead long and healthy lives unless they are subsidized by people offering them food and medical attention. When people do provide outdoor cats with medical treatment and care, or even just supplemental feeding, these felines can survive and even thrive, and may decimate local bird populations. They may be especially dangerous for migrating birds passing through unfamiliar areas. Outdoor and feral cats quickly figure out the patterns of when and where new migrants arrive, but these birds have no way of anticipating the presence of cats until it's too late.

As a former rehabber, I have cared for hundreds of birds that had been attacked by cats. In all but one case, these birds have died from internal injuries or infections from the puncture bites. The only cat-injured bird I've ever restored to health required antibiotics for three full weeks.

In my lifetime I've taken in five stray cats. They all adapted well to indoor life. It's harder to confine a cat that has spent its whole life going in and out but it's not impossible. If you feel you must allow your cat to go outdoors, you can at least try to reduce the harm to birds by letting it out only at night.

ˇ ˇ

Q I was thrilled when chickadees laid eggs in our nest box, but a predator got them before they hatched. How do I keep raccoons, cats, snakes, and other predators out of my birdhouses?

A A wide variety of commercial baffles and predator guards are available at bird specialty stores or on the Internet, or you can fashion your own. Some are designed to keep critters from climbing up the pole or tree, others to keep them from entering or reaching into the entrance hole. Unfortunately, some predator guards cause more problems than they solve, so be careful that your birds don't have trouble getting in or feeding their young with these guards in place. Because predators vary locally, seek advice whenever possible from others in your community who have birdhouses and who may know about solutions that work in your area.

SOLVING THE FERAL CAT PROBLEM ONE CAT AT A TIME

My daughter Katie and her college roommate Stacey had a problem: a feral cat was eating birds in their Ohio backyard. The cat was young, beautiful, and hungry, but so were the birds she was eating.

When I visited, I witnessed the cat killing a beautiful Carolina Wren — an adult male who was helping his mate feed nestlings at the time. Without his help, the female wren would have to work much harder to raise their young successfully. The last straw was when I watched the cat stalking the mother wren.

But what to do? I headed to the grocery store and bought a can of cat food, using it to entice her into my car. I took an experimental drive around the block with the cat eating in the back seat. When we got back and I opened the car door, I expected her to make a run for it, but she continued washing her paws in satiated contentment in the back seat. So I drove 800 miles back to Minnesota with her.

"Kasey" was apparently so hungry for regular meals and a home to call her own that she was perfectly happy and well behaved in my car. She was infested with worms and lice, but after a few relatively small vet bills she became a healthy, happy indoor cat and my treasured companion.

Q Recently, I have been seeing cowbirds in my yard. I wish they would go away because I've heard that they lay their eggs in other birds' nests. What can I do about it?

A It's true that cowbirds lay their eggs in the nests of other birds that raise the hatchling cowbird, often at the expense of at least one or two of their own young. Cowbirds inhabit open grassland areas or areas near the edge of forests, and their range and their numbers in many areas have expanded as humans cut down forests for agriculture and development. In some wildlife refuges and other areas managed for critically endangered songbirds, cowbirds are legally trapped and humanely euthanized. But this isn't permitted in most areas. Cowbirds are native American birds, covered by the same legal protections as robins and hummingbirds. It's important to remember that cowbirds are fascinating birds in their own right and can only survive as nest parasites.

The most important thing you can do is to stop subsidizing them. Close down feeders that they visit or switch to foods they don't like as much, such as striped sunflower or safflower seed. On a wider scale, encourage your local, county, and regional planners to limit the fragmentation of forests. We don't have much power in most zoning and development situations, and even when our yards are covered with native vegetation rather than turf, roads and driveways create enough openings for cowbirds to feel welcome. So they're an exceptionally frustrating problem. I wish I had a magic solution!

SEE ALSO: *pages 151–155 for more about cowbird behavior.*

Q I was excited to see a hawk in my yard a few weeks ago, until it started attacking the birds at my feeder. I even watched it eat one of my doves! What should I do?

A A few hawks, most often Cooper's Hawks and, in areas of Canada and around the Great Lakes in the United States, Merlins (noisy, small falcons) have adapted to nesting in residential neighborhoods and have discovered that feeding stations are a reliable place to find easy prey. If you don't want your backyard to be a regular hunting ground for them, it's a good idea to close down your feeders for a few weeks so the raptors can develop a different routine. If they nest in your yard or a neighbor's, you may want to keep your feeders closed for the summer and enjoy the opportunity to watch the nesting habits of these interesting raptors. But remember that hawks are birds, too. It's hard for them to understand that a bird feeder isn't a place for birds to feed on birds.

⌄ ⌄

Q A Baltimore Oriole has taken over my hummingbird feeder and won't let the little guys eat. What should I do to help my hummers?

A Like hummingbirds, many orioles feed on nectar. Orioles are considerably larger — a Baltimore Oriole is nine times heavier than a Ruby-throated Hummingbird, so if the oriole is the least bit territorial around the feeder, the hummers will hold back. But hummingbirds are probably the most aggressively territorial feeder birds, so even if your oriole were to leave, the hummingbirds would still fight over your feeder.

Probably the best solution would be to purchase a few more small hummingbird feeders to set in other areas of your yard. If you choose models without perches, the oriole won't be able to use those, leaving them entirely to the hummingbirds. And by setting out two or three, your hummingbirds will be able to spend more time feeding and less time squabbling among themselves. You'll be able to enjoy two of our most colorful, beautiful species!

∧∨

Helping Birds in Trouble

Q **A hummingbird got stuck in my neighbor's garage and died before she could get it out. What should I do if this happens at my house?**

A When birds panic, they tend to fly upward where, in their natural environment, they can escape every danger except aerial predators. So don't chase any bird in the garage. Chances are high that it will stress the bird and cause it to stay up too high, where it can't find the way out.

First, of course, open the door and windows, and cover with newspapers any windows that can't be opened. Then try hanging something brilliant red — a hummingbird feeder is the best choice — right at the entrance.

Hummingbirds are often attracted into garages in the first place by the bright red door pulls suspended from many garage doors. If you have a recurring problem with hummingbirds flying in, try covering the door pull with black electrical tape.

Sometimes birds enter homes through windows or open doors. When a bird is trapped in the house, it's best to try to confine it to a single room, closing all the doors. If you can completely open the windows in that room, sometimes the bird will fly out rather easily. It's most effective if the windows have shades that can be drawn over the glass so the only visible part of the window is wide open to the outdoors. If you can't open the windows, then close the drapes to darken the room as much as possible and, moving as quietly as you can, sneak up on the bird and toss a light towel over it. Then scoop it up, take it outside, and release it.

˅ ˅

Q **I found a pigeon with a band on its leg. It's very tame and came right into the garage. What should I do?**

A This bird is probably a homing pigeon. This breed belongs to the same species as common city pigeons (Rock Pigeons), but has been raised in captivity, so it trusts and depends on people. Like city pigeons, racing pigeons can be a variety of colors. Pure white homing pigeons are often raised for "dove releases," to be let go at a wedding or other event. All of these homing pigeons usually find their way home directly.

But sometimes a homing pigeon gets lost or exhausted on a long journey, and rarely one gets injured. These birds often turn to strangers, as yours did, for food, water, and shelter.

Tracking down the bird's owner can be a genuine kindness, both for the bird and the owner. Homing pigeons never wear the U.S. Fish and Wildlife Service bands used on wild birds. You can locate the owner through the pigeon racing organization that produced the band, by looking for the letter code on the band. In North America, most bands have one of these codes: AU (American Racing Pigeon Union), CU (Canadian Racing Pigeon Union), IPB (Independent Pigeon Breeders), IF (International Federation of American Pigeon Fanciers), NBRC (National Birmingham Roller Club), or NPA (National Pigeon Association). These organizations' websites will give you the information you need to find the owner.

✓ ✓

Q **What should I do if a bird crashes into my window?**

A About half of the birds that initially survive a window collision end up dying from their injuries. They may be captured by predators while they are stunned, or succumb to broken wing bones, bruises, and serious internal injuries. But about half do survive.

If a bird gets knocked to the ground and lands upside down, it may lie there, unable to right itself for many minutes, and the entire time it's on the ground it's vulnerable to predators; so the best thing is to go outside and pick it up. That may be all the attention it needs — it may fly off instantly. If it pecks at you but can't fly away, it is probably fairly seriously injured and should be brought to a wildlife rehabilitator immediately. If it is rather lethargic and can't fly, it may have a minor concussion but may still be able to perch and balance. If that's the case, place it in a nearby shrub and leave it to its own devices.

If the bird can't balance or perch, place it in a shoebox lined with paper towels and bring it inside. You can help it stay upright by fashioning a donut cushion from tissues. In winter, keep a bird with extremely thick feathers in a basement or other fairly cool but not cold place rather than in a room that will feel excessively hot to it. Every 15 minutes or so, take the box outside and open it, to see if the bird flies off. Don't open the box indoors! If the bird doesn't recover within an hour or two, bring it to a rehabber. Never release any songbird at nighttime. It won't be able to see well enough to find a safe roosting place.

SEE ALSO: *pages 90–91 for information about preventing window collisions.*

A HELPING HAND

To find a wildlife rehabber in your area, check the online directory at *www.tc.umn.edu/~devo0028/contact.htm.*

Q We have a sparrow near our feeder that is just sitting still and not eating much. I'm afraid that it's sick. What should we do?

A If a bird in your yard seems lethargic, is sitting still with puffed-out feathers, has crusted eyes, or shows other signs of illness, it's best to immediately bring your feeders inside, clean and air-dry them thoroughly, and don't begin feeding again for a week or so. This won't help your sick bird, but it will send other birds away to reduce the chance of the disease spreading to them. If you have reason to believe that the bird was stunned by striking a window or was injured by a cat, then there's no reason to close down your feeding station.

Don't catch a possibly sick bird unless you already have discussed it with a wildlife rehabilitation clinic and have been told specifically that they will be able to take it. Many rehabbers are reluctant to take sick birds because they don't want to put birds they're already caring for at risk of communicable diseases. And unless you're qualified and licensed, you may very well be letting yourself in for problems if you try to take care of it yourself. When birds are feeling sick or weak, they seldom preen, and lice and mites multiply quickly.

∨ ∨

Q When I was taking a walk with a friend, we came upon an injured bird. We were both afraid to pick it up, so we just kept going. What should we have done?

A If a bird may be sick, it's best to leave handling it to a wildlife rehabber. Depending on the disease possibilities, it may pose serious dangers to other birds at a rehab facility, and in some rare cases, the illness may even spread to humans, so interfering could cause worse problems.

If you can't bear to leave a bird, or if it's injured with little likelihood of making you sick (broken wings and cat bites are not contagious), you can help by transporting it to a wildlife rehabilitator. First put it in a cardboard box or a paper bag. Wear gloves if at all possible — strong leather gloves if the bird has talons (any hawk or owl) or has a sharp, potentially dangerous beak.

I keep a cardboard box lined with Astroturf on the bottom to provide footing (a few layers of soft paper towels can serve) in my car for this kind of emergency. I have a few numbers of local rehab centers written on the outside of the box so I can call the nearest one if I do find an injured bird.

The American Crow appears to be the biggest victim of West Nile virus, a disease introduced to North America in the 1990s. Virtually every crow who is infected dies within one week. No other North American bird is dying at the same rate from the disease, and the loss of crows in some areas has been severe. After the disease dies out in an area, crow numbers do seem to slowly recover.

DISPOSING OF DEAD BIRDS

Diseases that have been in the news in the past decade include a few that affected birds, too. West Nile virus, avian flu, and some salmonella outbreaks have raised everyone's awareness and concern when they see a dead bird. Dead birds are sometimes of interest to health officials and scientists.

If you're aware of a disease outbreak or are concerned about health issues, contact your local or county health department or the U.S. Geological Survey's National Wildlife Health Center *(www.nwhc.usgs.gov)*. Collect or dispose of the dead bird as they direct you. In many cases health departments will not be able to analyze a bird that has already started to decay, so you may be asked to double-bag it and put it in your freezer, or to take it to them immediately.

If you do pick up the bird be sure to wear disposable gloves or insert your hand into a plastic bag, pick up the bird with that hand, and then turn the bag inside out to contain the bird. Even if you're pretty sure your skin didn't touch the bird, wash your hands thoroughly afterward.

After any health and safety issues have been resolved, and especially if you know this bird was killed by a cat or in a collision with a window or automobile, or in some other way not associated with disease, you might turn your thoughts to collecting the bird for scientists at a university or museum. Start by contacting a wildlife professional who has a federal and state permit to collect birds or bird parts. (You may find such a person at a nearby university, museum, nature center, or an elementary or high school.)

The Migratory Bird Treaty Act of 1918 protects native American birds, dead and alive, and their parts (feathers, eggs, and nests), by forbidding anyone without a permit to own or handle birds or bird parts. Though at first glance the law may seem overly strict, it serves an important conservation purpose by allowing authorities to curtail activities that harm birds. By having oral permission to salvage the dead bird, you'll be able to show that you weren't salvaging it to sell or possess.

If you're instructed to bring the bird in under the authority of someone else's permit, remember to record your name and contact information, the date and location, the bird's species (if known), and a description of the circumstances, including your best guess about the cause of the bird's death. Use a pencil or permanent ink. If you're instructed to freeze the bird until you can bring it to the facility, double-bag it in plastic, and put the paper with this information between the two layers.

Q **I noticed a House Finch at my feeder that looked like it was sick. When I looked with binoculars, I saw that one of its eyes was all swollen and gunked up. What was wrong with it?**

A That bird was suffering from a nasty form of conjunctivitis called House Finch Eye Disease. The disease, caused by *Mycoplasma gallisepticum*, a common pathogen in domestic turkeys and chickens, had not been reported in songbirds until an outbreak in House Finches was first noticed during the winter of 1993–94 in Virginia and Maryland. The disease isn't harmful to humans, but it can be fatal to House Finches. Volunteer birdwatchers joined the Cornell Lab of Ornithology's House Finch Disease Survey to help track the spread of the disease across the continent. Their reports helped scientists understand the dynamics of epidemics in birds. To prevent this sick bird from infecting other birds at your feeders, you should close down your feeding station for a couple of weeks. Thoroughly wash your feeders and let them air dry until you put them up again.

❧ ❧

Q **I was cleaning out my nest boxes and I found a dead adult Tree Swallow in one of them. How did it die?**

A Tree Swallows migrate a long way — some of the birds that nest in northern Canada and Alaska winter down in Central America. If they arrive when the temperature is too cold for flying insects, their primary food, they may die of starvation or hypothermia. This is probably what happened to your swallow, especially if swallows or bluebirds have used your nest box successfully in past years.

There are a few other reasons why adult birds may die in a nest box.

▶ If it had obvious injuries, especially on its head, it may have been killed by a House Sparrow trying to take over the box. In most cases, though, these competitors toss out birds after they kill them. Find ways to exclude House Sparrows from your nest boxes at *www.sialis.org/hosp.htm*.

▶ Was the inside front of the box, below the hole, rough or grooved? Sometimes birds get stuck inside boxes because the inside walls are so smooth they can't climb out. Tacking sandpaper or small strips of wood, making a sort of ladder, will prevent this in the future.

▶ Sometimes an infestation of blowflies or other parasites can become so intense that it kills not only nestlings but also adults. If there was no sign of young, that's probably not the answer in this case.

▶ Some wood preservatives may release harmful gases, especially in hot weather. Make sure any paints or varnishes that you use on your nest boxes are rated safe for indoor or playground use.

Nestwatch, a citizen-science project of the Cornell Lab of Ornithology, provides a wealth of information for people with nest boxes. Consider reporting to NestWatch on the successes or failures of your nesting birds, to help scientists understand more about our breeding birds.

❤ ❤

Q **A robin missing an eye has turned up in our yard. He seems to be eating and acting normally. Should we help him?**

A Many birds adapt to monocular vision and survive for years in the wild with one eye. When the eye is first injured, sometimes potentially dangerous infections set in. If the bird becomes lethargic and easy to catch, it probably is infected and should be captured and brought to a wildlife rehabilitator. (You can locate the nearest one to you at *www.tc.umn. edu/~devo0028/contact.htm.*) But as long as it's acting normally and staying well out of reach, it's doing fine and will be better off if left alone.

❤ ❤

Q **I saw a chickadee at my feeder with a bill so long and curved that it was having trouble eating. What was wrong with it?**

A A large number of chickadees and other birds in Alaska have been developing unusual bills, often overgrown or crossed, in the past two decades. The outer sheath of the bill is made of a type of keratin, much like your fingernails, and in these birds this protective sheath is growing abnormally.

Colleen Handel, a biologist with the USGS Alaska Science Center, has documented bill deformities among 30 species in Alaska, from ravens and magpies to chickadees and nuthatches, and compiled records of at least 2,100 individual deformed chickadees in Alaska between 1991 and 2009 and 420 individuals of other species since 1986. She has been actively soliciting reports of deformed birds from outside Alaska but has received only 30 reports of deformed chickadees and 110 reports of other species from the rest of North America. Alaska has the largest concentration of bill deformities ever documented in the world.

Blood tests on birds with deformed bills found damaged DNA, consistent with environmental contamination or disease, though there have been no other obvious indications of disease. The reports cluster in late winter. Birds manufacture their own vitamin D, as we do, from exposure to sunlight, and vitamin D helps us absorb calcium. So chickadees visiting feeders, eating a higher proportion of seeds than they do on a natural winter diet of insect eggs and pupae, may be vulnerable to calcium deficiencies during the time of year when sunlight is lowest. This doesn't explain the numbers of other, unrelated species that don't visit feeders but still exhibit deformed bills.

SEE ALSO: *pages 269, 271, and 276 for information about helping baby birds.*

CHAPTER 4

Looking to the Future: Protecting Bird Populations

In the twenty-first century, many birds are in trouble. Some of America's most treasured grassland birds, including Bobolinks, Eastern and Western meadowlarks, and Northern Bobwhites, are showing steep declines. Of the 71 bird species unique to Hawaii during historic times, 26 have become extinct, and 30 of the remaining species are threatened or endangered. More than 75 percent of birds nesting in America's aridlands, including sage-grouse, California Condors, and Elf Owls, are declining steeply or already listed as threatened or endangered.

Wetland bird populations are well below their historic levels, but many wetland species, from Bald Eagles and Osprey to American White Pelicans and Sandhill Cranes, are now thriving thanks to wetland restoration, which has become a model for bird conservation. Peregrine Falcons, extirpated from eastern North America by the 1970s, are now nesting in many areas after benefitting from the same protective measures that helped raptors in general, and also from reintroduction projects.

Strategic land management and conservation action are the tools we use to help species in peril. Our successes prove that when we set our minds to it, we can make the difference between life and death for populations and species.

^v^v^v^v^v^v^v^v^v^v^v^v^v^v^v^v^v^v^v^v^v^v^v^v^v^v^v^v^v^v^v^v

Make Way for Plovers

Q **Piping Plovers used to nest on the Lake Superior beach near my cabin. Last spring two showed up during migration, and the beach was closed for a while. But people ignored the signs and took their dogs on their regular walks, and the plovers moved on. What can we do to help these little birds?**

A Many people have never heard of Piping Plovers, and so naturally some might feel resentful if told to walk their dogs somewhere else. One strategy that has allowed the closely related Snowy Plover to make a comeback on a Santa Barbara beach involves introducing people to this splendid bird.

Piping Plovers are specialists, with adaptations that make them perfectly suited for life on beaches and mudflats but virtually nowhere else. They feed on tiny water creatures washed to shore or in waterlogged sand along shallow rivers, large lakes, and oceans. They pick at visible bugs and bring small invertebrates to the surface by extending one foot slightly forward and vibrating it against the wet sand.

Piping Plovers are exceptionally hardy for being so diminutive. They nest in a scrape on open sand, gravel, or shell-covered substrate, or in dunes along shorelines. Their tiny chicks are exposed to high winds and blowing sand that would send most birds scurrying. The eggs and adult and chick plumage are perfectly

camouflaged to protect them from Peregrine Falcons and other natural beach predators.

Though exquisitely adapted to this wild environment, Piping Plovers are not adapted to the changes that people bring. Where beaches haven't been developed, all-terrain vehicles, running dogs, oil slicks, and other disturbances make it difficult for plovers to feed and nest. Gulls, raccoons, and crows that are initially attracted to a beach by picnickers and their garbage will also notice and eat plover eggs and chicks. Feral cat colonies imperil them in some areas.

At the public beach on the University of California-Santa Barbara's Coal Oil Point Reserve, people wanted to protect the closely related and threatened Snowy Plover, which had stopped breeding along that beach. The U.S. Geological Survey (USGS), the Santa Barbara Audubon Society, and the University of California Natural Reserve System worked together to fashion a program that has brought the number of plover nests there annually from virtually zero from the 1970s through 2000 to a few dozen every year since 2004. How did they do this?

A USGS study determined the smallest part of the beach that could be closed to maximize protection of plovers with minimal inconvenience to beach users. In 2001, when a single Snowy Plover chick was seen near a dune-restoration project, the university installed a rope fence to enclose the core plover habitat along 400 yards of beach. The area stretched from wet sand to dry areas above the tidal zone but allowed people to walk at the water's edge along the beach. Educational and regulatory signs were installed.

Perhaps the single most critical thing the partnership did was to empower an army of volunteer docents to become Snowy Plover ambassadors. I went to the beach to see them in action in 2005. I witnessed their interactions with passersby, showing and educating them about the plovers, requesting compliance with the dog leash law, requesting them to keep their distance from the fence, and scaring away crows. A spotting scope was trained on a plover chick, and I loved hearing people seeing them for the first time: "They're so adorable!" "They look like marshmallows on stilts!" "How come I never heard of these birds before? They're wonderful!"

Making people aware of the plovers and their plight in such a positive way has made protecting them much easier.

v v v v v v v v v v v v v v v v v v v v v v v v v v

Q I own some forested property that I've been managing for income but also for the environment and to give something back to my local birds. I was going to plant some oaks but experienced foresters told me not to, because they'd never be big enough to bring in any income during my lifetime. What do you think?

A Planting trees is something we do for the future. Even young oaks can provide a lot of benefits for migrating Scarlet Tanagers and other birds, and for your viewing pleasure. My recommendation is to find out from your county extension office or state department of natural resources what the dominant forest type was in your area before settlement, and then choose a variety of locally native species to plant. Select a mixture that will allow you to selectively harvest some trees soon, for profit, while allowing other varieties to grow for the future benefit of birds and your descendents.

⌄ ⌄

Q **Our city recently passed an ordinance that allows people to replace lawn with native plants. Is this just a fad? Won't it be bad for robins?**

A Robins do love taking worms from lawns! But robins cannot live by worms alone — a large part of their diet comes from insects and fruits associated with other kinds of vegetation. Compared to turf lawns, natural plantings provide a much wider variety of animals, most notably butterflies and birds, with both food and shelter, while still fostering worms for robins to enjoy. And locally native plants are adapted to locally native conditions, requiring less watering, fertilizing, and pesticide applications than turf, allowing us to create a safer environment for people and birds.

Turf lawns are a necessary element in golf courses and playing fields, and will always appeal to some homeowners as well. But when local ordinances allow homeowners to replace traditional lawns with natural vegetation, they're acknowledging the different aesthetic tastes of constituents. And allowing natural vegetation also acknowledges the different tastes of the avian community, from robins to hummingbirds.

THE CANARY IN THE COAL MINE: DEALING WITH PESTICIDES

Miners have long brought canaries into mines, knowing that the death of a canary was a warning of a clear and present danger to them, too. Wild birds have provided similar warnings to us from above ground. In the spring of 1955, George J. Wallace, professor of ornithology at Michigan State University, noticed that robins were dying around campus, and connected it to insecticides. By the summer of 1958, robins had been eliminated from campus and parts of the surrounding city.

Wallace's work documented the devastating effect that DDT was having on songbirds. Meanwhile, Peregrine Falcons, Bald Eagles, and Osprey were declining dramatically, but weren't dying outright. However, virtually no juvenile birds were being seen anymore. In 1968, Daniel Anderson and Joseph Hickey published a paper in *Science* documenting eggshell thinning in these birds, coinciding with the introduction of pesticides like DDT.

DDT killed insects on food crops and mosquitoes carrying malaria, so many argued that human beings should take precedence over birds. Although the deaths of songbirds and eggshell thinning in raptors did figure in Congress's decision to ban DDT in 1972, another factor was the discovery of DDT and its byproducts in human mother's milk.

The decision to ban DDT proved to have saved human lives. In 2002, scientists analyzing stored blood samples of pregnant women from the 1960s discovered that DDT levels in blood serum were highly correlated with low birth rates and premature births. Those "canaries in the coal mine" were giving us a sound warning. Meanwhile, in every place where DDT has been applied in the outdoor environment, mosquitoes were growing ever more resistant to it.

Although DDT is banned in the United States and other developed countries, in areas of the world where malaria is a problem, the World Health Organization today recommends the use of indoor spraying. DDT is taken up in the feet of mosquitoes, so when ceilings and bed netting are sprayed, they afford protection while minimizing effects on our own and natural food chains.

In *Silent Spring,* Rachel Carson wanted us to be mindful of the negative as well as positive effects of pesticides, to seek out alternatives when feasible, and when a pesticide seems to be the best solution to a serious problem, to use the smallest effective amount — a common sense approach for protecting humans and our crops as well as the natural environment.

Q A couple of times this year after my neighbor had her lawn sprayed, I found robins acting strangely, acting sick and even falling on their sides or on their backs. They ended up dying. My neighbor says that the pesticides she used are approved by the EPA and so are guaranteed not to hurt birds. Is that true?

A The EPA never "approves" pesticides; it registers them. And field-testing to prove a pesticide is safe for wildlife is no longer a requirement for registration; in order to be registered, a product must pass a complicated cost/benefit test; no tests on birds are required. Even a major bird-killing pesticide such as fenthion can remain on the market for many years, killing millions of birds, before its use is restricted or prohibited.

According to the U.S. Fish and Wildlife Service, about seven million birds are killed every year by common household lawn pesticides. In general, insecticides are more dangerous than herbicides, but both can cause neurological damage and cancers in birds and even people. It's difficult and expensive to necropsy and do the necessary tests to learn if pesticides have killed a bird. Most modern pesticides break down very quickly, both in the environment and in the body, so finding the definitive cause of your robins would be difficult and expensive at best.

In addition to weed killers for destroying dandelions and insecticides for killing cutworms, some lawn care applications include fertilizers. Although fertilizers would not be to blame for sick robins, they seep down into groundwater, eventually working their way into lakes, rivers, and streams, where they contribute to the excessive growth of aquatic plants. So it's very important to minimize the use of fertilizers as well as pesticides. Be sure to check the labels of every product you use, and if you hire a company to apply lawn products, make sure they tell you what products they're using and the purpose of each one. In all cases, it's sensible to apply the absolute minimum amount that would be effective for the problem a homeowner is tackling. Unless weeds are dire, it's much wiser to pull or spot spray weeds than to spray every inch of lawn.

Better yet, an integrated pest management system can often maintain beautiful yards without formulated chemicals. The EPA has information at *www.epa.gov/opp00001/factsheets/ipm. htm* about using this "common-sense, effective, and environmentally sensitive approach to pest management" that will help you keep your family, pets, and wild birds safe. If your neighbors see your success with it, perhaps they'll consider trying it, too.

˅ ˅

Q **When my nephew graduated from elementary school, I wanted to have a big balloon release but he said that would be bad for birds and, of all things, turtles! Is this true?**

A Balloons are lovely floating in the sky, but eventually they come back to earth. Since two-thirds of our planet is covered by water, a great many released balloons find their way to lakes and oceans. Marine mammals, sea turtles, and many seabirds such as pelicans are killed or seriously injured when they encounter them and get entangled in the strings or swallow them. Strings from balloons sometimes ensnare birds in trees, too. You're lucky to have such a conscientious and aware nephew! He may appreciate a tree planted in his honor that will provide food and shelter for the birds he treasures for years to come. (Of course, he'll also appreciate a gift he can use right now, too!)

~v~v~v~v~v~v~v~v~v~v~v~v~v~v~v~v~v~v~v~v~v~v~v~v~v~v~v~

The Effects of the Environment on Birds

Q **How are birds affected by climate change?**

A Climate change is influencing the abundance, distribution, and timing of migration and breeding for many species. So far various climactic changes have helped some species, harmed others, and had little effect on the rest. A recent study by the National Audubon Society showed that more than half of the birds commonly found on the Christmas Bird Count are wintering farther north now than they did 40 years ago.

American Robins are now arriving approximately 14 days earlier than they did in 1981 on their breeding grounds in the Colorado Rocky Mountains. Tree Swallows have advanced their breeding date by up to 9 days earlier from 1959 to 1994. Red-winged Blackbirds, Eastern Bluebirds, and eastern populations of Song Sparrows now lay their eggs earlier because spring temperatures are warmer.

In addition to these effects on migration and breeding, birds are at risk from habitat changes caused by climate change, especially on the tundra, in alpine meadows, on sea ice and glaciers, in coastal wetlands, marine atolls, and oceans.

These species may face severe conservation challenges in the coming decades. Sea level rise may inundate islands, jeopardizing nesting birds. The potential spread of mosquito-borne avian malaria to highlands where the surviving Hawaiian honeycreepers have retreated is also a danger.

In the future, climate change is expected to affect the survival and reproduction of many bird species. Again, changes are expected to benefit some species and harm others. Changes in prey distribution and abundance, shrinking habitats, and changes in rainfall and water availability are expected to present great challenges to some birds on land and at sea.

Eastern Bluebirds occur across eastern North America and south as far as Nicaragua. Bluebirds living in the more northern and western parts of the range usually lay more eggs than do the more eastern and southern birds. Eastern Bluebirds typically have more than one successful brood per year. Young produced in early nests usually leave their parents during their first summer, but young from later nests frequently stay with their parents through the winter.

WEATHER: THE GOOD, THE BAD, AND THE UGLY

Only in cartoons do birds lounge atop rainbow colored clouds. In the real world, weather events may prove beneficial to some birds and harmful to others. Overall, birds are best adapted to the normal weather patterns of the areas where they have evolved, but birds use their wings to move about and are constantly developing new adaptations to new conditions.

Sudden Storms

For some birds, the more rain, the better. The nesting success of Song Sparrows is directly related to precipitation levels, but cool, rainy springs coincide with lower nesting success for Tree and Cliff swallows.

Storm systems are, and have always been, dangerous for many birds. In the early 1800s, John James Audubon watched as two nighthawks were struck and killed by lightning. In 1941, a lightning bolt blasted four Double-crested Cormorants from a flock. All four died, though their feathers weren't scorched. That same year, more than 50 Snow Geese were struck by a single lightning bolt. Necropsies revealed that one was badly burned, but most had died from the impact as they hit the ground.

Many trees felled by lightning have been found to contain dead woodpeckers, owls, or other birds. In 1953, two hailstorms in Alberta killed 150,000 ducks and geese. In 1960, thousands of Sandhill cranes were killed by hail in New Mexico. In 1931, a bluebird pelted by hail in Iowa suffered two broken wings. In 1938, hail killed two California Condors that had been eating a horse carcass. In May 2004, a hail storm killed more than 100 nesting

Great Blue Herons in northern Wisconsin. Yet even in the face of that devastating storm, at least 50 herons were unharmed.

Hurricanes can kill birds outright. They can also devastate the vegetation along huge swaths of shoreline. And because so many houses, garages, industrial plants, gas stations, sewage facilities, and other human structures can get flooded, toxins and debris can be released into floodwaters, degrading coastal wetlands even more.

Devastating Droughts

Droughts and floods have always played a part in the natural world. Although severe ones can devastate local populations of birds, on a wider scale they've been balanced out by favorable conditions elsewhere. If climate change alters rainfall patterns so some areas become consistently drier or moister, this will change the composition of vegetation and insect life of those areas, increasing bird populations of species that associate with increasing plants or insects and decreasing bird populations of those that don't.

Droughts can have unexpected impacts on birds. For example, when water levels drop, feeding swans often pick up grit at the bottom of shallow waters that are normally too deep for them to reach. Most North American lakes have large amounts of lead shot sitting at the bottom like tiny time bombs. Efforts to phase out lead shot began in the 1970s, but a nationwide U.S. ban on lead shot for all waterfowl hunting was not implemented until 1991. Canada instituted a complete ban on the use of lead shot in 1999, after banning its use near bodies of water and on national wildlife areas earlier. Even though lead shot is no longer

129

raining down on lakes, rivers, and streams, the lead shot already there continues to sit until a swan, loon, or other bird picks it up as grit. The lead is ground up in the gizzard, dissolves into the bloodstream, and usually leads to the bird's death.

Rising Temperatures

Within certain ranges, insects are more active as temperatures rise, so birds depending on them may benefit from a warming trend. Considering how many birds are found in tropical areas compared to temperate ones (tiny Costa Rica, about the size of West Virginia, has more bird species than all of North America north of Mexico!), it's possible that in the long run, in some places, birds may benefit from rising temperatures.

But harmful insects would also grow more active. In Hawaii, where native birds were decimated by malaria after mosquitoes were introduced, most remaining populations of several endangered species now exist only at elevations too cool for mosquitoes. As temperatures rise, mosquitoes and the diseases they carry will move higher in the mountains, shrinking and even eliminating remaining habitat for these vulnerable birds.

Rising water temperatures also encourage *eutrophication*, marked by excessive aquatic plant growth. While the plants are actively growing, they produce oxygen, but as they die, even as other plants are growing, decomposition of dead plants absorbs oxygen and releases carbon dioxide. Mosquito larvae can thrive in oxygen-depleted water because they take in their oxygen at the surface through a breathing tube.

Mayfly nymphs feed on mosquito and other larvae in their aquatic stage, and provide enormous quantities of food for migrating and nesting birds when they emerge as flying insects in their adult stage. But mayflies require highly oxygenated water. When eutrophication makes lakes and ponds inhospitable for them, we lose a natural control over mosquito numbers and an important supply of food for such beloved and already-declining birds as Purple Martins and Common Nighthawks.

As warmer temperatures arrive earlier in spring, a mismatch between timing of nesting and availability of food seems to be arising. For example, many plants and insects are emerging earlier. Migratory songbirds that spend the winter in the southern and central states may take advantage of warm temperatures and earlier food availability to migrate and breed earlier. Meanwhile, birds wintering in the tropics have no way of knowing what the weather is like across the Gulf of Mexico, and arrive closer to their normal arrival date, sometimes missing important surges in insect numbers that fuel their migration and provide a good abundance of food for breeding.

Finally, the northern range limits of some opportunistic birds are related to temperature. When there is less ice in winter, Mallards and Canada Geese winter farther north, and their winter range may expand. More and more starlings, robins, and other abundant species may survive in higher numbers in northern winters, in some cases crowding out birds that are more adapted to harsh conditions; as those populations retreat further north, their potential total range will shrink. Melting permafrost is expected to affect lemmings, the primary food for Snowy Owls. Many species of birds are physiologically stressed by extreme heat, too; increasingly hot summers may take a heavy toll on these species.

Q I keep hearing about birds in decline, but I'm seeing more robins than ever. What's the truth?

A You're right that robins are increasing. These *generalists* have adapted very well to the habitat changes that come with urbanization, and they are thriving. But some birds are dangerously declining. In 2007, the National Audubon Society released a report that showed many of our most common and beloved birds have declined by 50 percent or more since 1967. For example, Eastern Meadowlarks have declined by 72 percent.

The population density and range of Greater Sage-Grouse have shrunk since Lewis and Clark documented them on their expedition. Common Nighthawks and Whip-poor-wills grow increasingly scarce. The Red Knot declined by fully 80 percent between 1995 and 2005, so in 2005, environmental organizations petitioned the U.S. Fish and Wildlife Service (USFWS) to list the *rufa* subspecies as Endangered and to designate critical habitat for them, but their petition was denied. It grows increasingly difficult to "list" declining species to give them

the protection that official designation as a Threatened or Endangered species affords.

Despite the declines of some of these species, there is plenty of room for hope. Habitat restoration has helped Kirtland's Warbler populations increase more than ten-fold in just three decades. Where wetlands are restored, waterfowl populations respond quickly. The Conservation Reserve Program, conservation easements, and other initiatives have helped declining grassland species, too. When we set our collective will to solving a problem, we usually succeed.

˅ ˅

Q When I was on a local bird walk in Florida, I saw some beautiful birds that the leader identified as Eurasian Collared-Doves. I was thrilled, but she said that they're not native to America and so are not a "good" species! Is that true?

A Eurasian Collared-Doves are beautiful birds and have an important place in the ecosystems of southeastern Europe and Japan, where they are native, but they are not native to Florida. These doves were released in the Bahamas by humans in the 1970s and had spread to Florida by 1982. They are still most abundant near the Gulf Coast but have spread as far as California, British Columbia, the Great Lakes region, and Veracruz in Mexico.

So far, Eurasian Collared-Doves in America don't seem to be competing with native species for food or nesting sites, and this introduction may well prove to be less destructive than most. But some introduced species are implicated in the declines of native birds because they compete for nesting sites. European Starlings and House Sparrows aggressively take over nesting cavities of native birds such as Red-headed Woodpeckers, Purple Martins, and bluebirds.

˅ ˅

Q Don't introduced birds contribute to an area's bio-diversity? And isn't that a good thing?

A Biodiversity is the variation of life forms within and among ecosystems and is often used as a measure of the health of a biological system. So it certainly does appear that by adding a new species, we're adding to biodiversity.

Sadly, it's not as simple as that. Consider invasive plants: when one — such as kudzu, purple loosestrife, Eurasian water milfoil, or cheatgrass — colonizes an area, it quickly crowds out a variety of native plants, greatly reducing the overall plant biodiversity of that area. That, in turn, reduces the biodiversity of animals. For example, when cheatgrass invades rangeland in the West, it crowds out native sage grass, and so is one of the primary causes of the dramatic decline of Greater Sage-Grouse.

Invasive nonnative birds can do the same thing. House Sparrows, like native bluebirds, wrens, Purple Martins, and Pro-thonotary Warblers, nest in cavities that they cannot excavate themselves. All of these species normally take over an aban-doned woodpecker hole, a cavity formed when a branch rots out, a nest box, or similar enclosed space. These cavities are, for many species, a *limiting factor.* When House Sparrows colonize an area, they aggressively evict native species from these nest sites, often destroying eggs and killing chicks and adults. Widespread colonization of House Sparrows is one of the factors that contributed to the decline of Purple Martins.

It's hard to predict which species will become invasive in a new place, and even harder to control an invasive species once it's released. Preventing introductions in the first place is vital for preserving biodiversity.

˅ ˅

Q Robins and Cedar Waxwings are always visiting my buckthorn trees. My mother-in-law says buckthorn is an "invasive exotic" and that I should get rid of it. How can it be bad if the birds love it?

A Common or purging buckthorn is indeed an invasive exotic in America. It was introduced here partly as a garden shrub and partly because of its purgative properties in herbal medicine. It's an invasive pest, crowding out other shrubs and serving as a primary host of the soybean aphid.

Birds do gorge on the berries, and then fly off to poop out the seeds in other places. This seems like a good deal for the birds, but buckthorn may not be very good for them and is definitely not good for the environment. Because buckthorn crowds out native shrubs, it reduces the amount of food available to birds in times of the year when it isn't fruiting.

So, yes, if common buckthorn is growing in your yard, it's a good idea to cut it down. It grows back quickly, so many authorities recommend that after you cut it down you treat the stump with an herbicide. If you do this, be careful to limit the application to only the buckthorn. Replace buckthorn with fruiting trees and shrubs native to your area. You can find out good varieties to choose from local conservation and gardening clubs or your state department of natural resources or environmental conservation.

Many birds that eat a lot of fruit have digestive tracts capable of separating out the seeds and regurgitating them, but the Cedar Waxwing lets the seeds pass right through. Scientists have used this trait to estimate how quickly waxwings can digest fruits.

TROUBLE IN PARADISE

🐦 **Invasive species can pose dangers** that aren't appreciated until it's too late. For example, the introduction of mosquitoes onto the Hawaiian Islands posed a serious problem for native humans and birds, but that problem was gravely exacerbated when nonnative birds were introduced. These birds from other areas carried blood-borne pathogens that they had long ago evolved immunities against. When mosquitoes bit them and then bit a native Hawaiian human or bird, they transferred these pathogens to them.

Malaria in particular decimated both the native human and bird populations of the Islands. Most native Hawaiian birds have been extirpated at the lower elevations of the islands where mosquitoes occur. One of the concerns about climate change on Hawaii is that warmer temperatures allow mosquitoes to spread to higher elevations, reducing the areas where native birds can still survive.

Many of the introduced birds on the Hawaiian Islands are stunningly beautiful, but in every case, they live in other places as well. Native Hawaiian birds, such as the Iiwi and Apapane, are found nowhere else on the planet. Of the 71 known endemic Hawaiian birds, 26 are extinct, and 30 of the remaining species and subspecies are listed as endangered or threatened by the U.S. Fish and Wildlife Service.

Extinction Is Forever

Q **I know about the Dodo and the Passenger Pigeon, but what other birds have gone extinct in the past century?**

A The last captive Carolina Parakeet died in a zoo in 1918. Eskimo Curlews may be extinct — coordinated efforts since the mid-1980s to locate these elegant shorebirds have failed, but isolated unconfirmed sightings even in the 2000s continue to surface so there is hope that a few still exist.

Many researchers believed Ivory-billed Woodpeckers had become extinct until 2004, when documented (but controversial) sightings were made of at least one male in the Big Woods of Arkansas. Intensive efforts to locate this species in Florida, Arkansas, and elsewhere have proven difficult, and no unequivocal proof that the species still exists has emerged.

Bachman's Warbler from the southeastern United States is most likely extinct. The last confirmed sightings in the United States were all near Charleston, South Carolina, in 1958–61, but scattered unconfirmed sightings in the Southeast and in Cuba continued into the 1980s.

Most American extinctions have taken place on Hawaii, where just during the twentieth century we lost the Laysan Rail, Hawaii Oo, Kauai Oo, Oahu, Olomao (probably), Lanai Hookbill, Greater Amakihi, Nukupuu (probably), Lesser Akialoa, Greater Akialoa, Kakawahie, Oahu Alauahio, Ula-ai-hawane, Black Mamo, and the Poo-uli, which was just declared extinct in 2004.

HELPING ENDANGERED SPECIES
MAKE A COMEBACK

When severely declining birds are listed as Threatened or
Endangered, the Endangered Species Act is extremely effective
at helping them. Bird conservation initiatives that were started
when the Endangered Species Act was enacted and enforced
have had some genuine successes. Peregrine Falcons, for example, had been extirpated from virtually the entire eastern half
of North America, but reintroductions, often using the chicks of
captive falcons from falconers, brought the species back so that
its status has been upgraded from Endangered to Threatened.

Bringing Back Raptors

After DDT was banned, Osprey and Bald Eagles made remarkable comebacks without reintroductions. During the 1960s, Bald
Eagle numbers remained fairly high in Alaska and much of Canada but declined to 417 breeding pairs in the Lower 48 by 1963.
By 1999, their numbers there had increased to more than 5,000
pairs. They have been removed from the Endangered species
list except in the Southwest region where they are designated
Threatened.

Between the 1950s and the 1970s, nesting Osprey declined
by 90 percent on the coast between New York City and Boston.
Their recovery has been startling — their 2001 population was
estimated to reach 16,000 to 19,000 pairs.

(continued)

Saving a Songbird

In 1975, surveys found fewer than 200 singing Kirtland's Warblers, limited to a small area in Michigan. The 2007 surveys found 1,707, including 8 in Wisconsin and 2 in Ontario.

Kirtland's Warblers are extraordinary specialists, nesting only on the ground beneath the bottom branches of jack pine trees. Jack pine is fire-adapted. Its cones remain tightly closed, often for many years, until exposed to the intense heat of fire, when the seeds are released and germinate on the burnt ground. As jack pine trees age, they lose their bottom branches, making them unsuitable for Kirtland's Warblers to nest. Once the jack pines in an area have lost their lower branches, that area will be abandoned by the warblers until fire starts the cycle over again.

Kirtland's Warblers were never abundant, but declined dangerously in the midtwentieth century until they numbered fewer than 200 pairs by the 1970s. Two factors caused their decline. Fire suppression prevented new jack pines from replacing old ones, and Brown-headed Cowbird populations mushroomed locally. By the late 1960s, in one sample of 29 nests, 70 percent were parasitized and only two fledgling warblers were produced.

Thanks to controlled burning and cowbird trapping, Kirtland's Warblers are currently estimated at over 3,000 individuals.

The Wonderful Whooper

Whooping Cranes, which had dwindled to 15 or 16 individuals in 1941, numbered 328 in the wild (including natural and introduced populations) as of April 2009. Protection of the wintering habitat at Aransas National Wildlife Refuge and intensive efforts at reintroduction are responsible for these increases.

Despite this success, Whooping Cranes are still extremely vulnerable. The natural population that winters in the estuary along the Texas Gulf Coast depends on rich supplies of blue crabs for their winter diet. When crab supplies are abundant, the birds have excellent winter survival and put on plenty of weight. During those years, they arrive on their spring breeding grounds in top condition, and their breeding success in the north is high.

But blue crab numbers decline as the estuary's salinity increases. More Whooping Cranes die during these winters, sometimes from emaciation, sometimes in collisions with power lines and other structures as they fly greater distances in search of food. The winter of 2008–09 was such a year; as of the end of April, they had lost 23 birds, more than an 8 percent loss.

The salinity of the marsh is weather-related; when rainfall declines, water levels go down in the Aransas River and other waterways emptying into the estuary. But these water levels are also related to human consumption of water, not just for drinking and agricultural irrigation but also for swimming pools and lawns. As the human population of Texas continues to grow, the future of the Whooping Crane remains in question. Fortunately, a great many people and organizations remain focused and committed to protecting this beautiful and charismatic species.

SEE ALSO: *pages 150, 213 and 235 for more on the Whooping Crane.*

Bringing Puffins Back

In 1885, hunters took the last Atlantic Puffins from Eastern Egg Rock in Maine's Muscongus Bay. By the midtwentieth century, puffins remained on only two islands in American waters, Machias Seal Island and Matinicus Rock, both off the coast of Maine.

Although puffins continued to do well on more northern islands, they didn't recolonize their historic nesting islands despite protection. Puffins tend to return to the islands where they were hatched, and once an island has no surviving birds, there are no birds to return to it.

In 1973, Dr. Stephen Kress started an experimental project, sponsored by the National Audubon Society, to return Atlantic Puffins to Eastern Egg Rock by raising young birds in burrows on the island. Between 1973 and 1986, 954 young puffins were transplanted from Newfoundland to Eastern Egg Rock. The young puffins were reared in artificial sod burrows for about one month. Audubon biologists placed handfuls of vitamin-fortified fish in their burrows each day, in effect taking the place of parent puffins, and 914 of these successfully fledged, flying to the ocean to spend the next few years. It takes a minimum of two to three years for young puffins to return to their islands to breed, and some of these puffins began returning to Eastern Egg Rock in June of 1977.

In 1984, National Audubon and the Canadian Wildlife Service began a similar puffin restoration project at Seal Island National Wildlife Refuge in outer Penobscot Bay. In 2008, thanks to Project Puffin, 101 puffin pairs were found nesting at Eastern Egg Rock and 375 pairs were counted at Seal Island NWR. The numbers are considered minimums, as Maine puffins nest in hidden burrows under boulders and are hard to count.

The restoration of Atlantic Puffins to U.S. coastal islands is a wonderful case study in how commitment and innovative action can bring a lost population back.

PART TWO

Bird Brains:
Avian Behavior
and Intelligence

CHAPTER 5

Actions Speak Louder in Birds: Bird Behavior

Birds are fascinating! Whether we're watching cardinals nesting in a shrub near our window, a Red-winged Blackbird chasing a huge Red-tailed Hawk in the sky, or a pair of dancing Western Grebes scooting across a lake on our TV screen, we can't help but marvel at the ways they go about their daily lives. Current research is establishing that birds are more intelligent than people once believed. Studies are also showing that birds such as chickadees and canaries replace brain neurons every year, allowing them to delete outdated memories and create new ones so their relatively small brain can keep up with their changing world. The more we learn, the more amazing birds prove to be.

^v^v^v^v^v^v^v^v^v^v^v^v^v^v^v^v^v^v^v^v^v^v^v^v^v^v^v^v^v^v^v

How to Think Like a Bird

Q How big are bird brains?

A Bird brains are 6 to 11 times larger than those of similarly sized reptiles — comparable to mammalian brains as a percentage of body mass. Because bird brains are somewhat similar in structure to reptilian ones, scientists have long labeled the structures using the same terminology as for reptile brains. But in 2005, a consortium of neuroscientists proposed renaming bird brain structures to portray birds as more comparable to mammals in their cognitive ability. The scientists asserted that the century-old traditional nomenclature is outdated and does not reflect new studies that reveal the brain-power of birds.

⌄ ⌄

Q I read in a newspaper column that hummingbirds remember the people who feed them. Is this possible?

A Alexander Skutch, who spent many decades studying birds in Costa Rica, wrote in his book *The Life of the Hummingbird* that after years of observation, he was convinced that they do remember. When a hummingbird comes to your window in spring and looks in as if expecting you to feed it, it's easy to jump to the conclusion that the bird is the same one you fed the previous summer, and it probably is. But we can't be absolutely certain about that without marking the bird in some way.

It's harder to systematically study hummingbirds than larger species; only a handful of people are licensed to band hummingbirds, and the birds are so tiny that leg bands aren't easy to read without recapturing the birds. Color-marking hummingbird feathers may affect how other birds respond to them, disrupting their normal social interactions and behaviors, and detailed research into the question hasn't been done. So although my guess is yes, we can't be scientifically certain.

CROWS NEVER FORGET EITHER

We know that American Crows remember human faces, thanks to a fascinating 2008 study. John M. Marzluff, a wildlife biologist at the University of Washington, noticed that birds he'd previously trapped seemed more wary of him and were harder to catch than crows that hadn't been trapped by him before. So he devised an experiment to see if the crows really could recognize his face.

To test whether the birds recognized faces separately from clothing, gait, and other characteristics, Dr. Marzluff got some rubber caveman and Dick Cheney masks. Wearing the caveman mask, he and his team trapped and banded seven crows on the university's campus in Seattle. In the months that followed, Dr. Marzluff and his students and volunteers walked prescribed routes around campus not bothering crows, but wearing either the caveman or the Dick Cheney mask. The crows scolded people in the "dangerous" caveman mask significantly more than they did before they were trapped, even when the mask was disguised with a hat or worn upside down. The Cheney mask provoked little reaction.

Over the following two years, even though no more crows were trapped or banded by people wearing the caveman mask, the effect actually multiplied. Wearing the caveman mask on one walk through campus, Dr. Marzluff was scolded by 47 of the 53 crows he encountered, many more than had experienced or witnessed the initial trapping. He hypothesizes that crows learn to recognize threatening humans from both experience and from their parents and others in their flock.

Then Dr. Marzluff repeated the experiment using more realistic masks made by a professional mask maker, trapping crows at various spots in Seattle while wearing one mask, the "dangerous" one. Afterward, he enlisted volunteers to walk through various areas of Seattle wearing either the dangerous mask or another style that the crows had not seen. The reaction to people wearing the dangerous mask was significant — crows scolded them, and in downtown Seattle even flew at them so close they almost touched them. The crows unerringly scolded the people wearing the dangerous mask rather than people wearing the mask that hadn't been worn during trapping.

Young American Crows typically do not breed until they are four years or older. In most populations the young help their parents raise new broods for a few years. Families may include up to 15 individuals and contain young from five different years.

Q I seem to remember that when I was a kid, I saw a TV special with a man doing a dance with a Whooping Crane. Did that really happen or was it my imagination?

A You're remembering a real bird and scientist whose wonderful interactions provided a lot of information about how to successfully rear some "imprinted" Whooping Cranes in captivity. Their bond actually led to a conservation triumph. The man was George Archibald, founder of the International Crane Foundation, and the Whooping Crane was Tex, a bird that had been hatched in the mid-1960s at the San Antonio Zoo and needed special care. Because she was hand-fed, she had imprinted on humans. Before she was old enough to breed, she was sent to the Patuxent Wildlife Center where they worked hard trying to get her to accept another Whooping Crane as a mate. But she clearly preferred her human handlers, never showed any interest in the other bird, and never laid an egg.

In 1975, Tex was moved to the International Crane Foundation in Wisconsin, under Archibald's care, and the scientist began a long-term experiment to see if this imprinted bird would form a pair bond with a human so she could be induced to ovulate and be artificially inseminated. He moved in with Tex for several months in 1976 and established a pair bond with her, regularly dancing with her. He followed Tex's lead and flapped and jumped. The next spring, she laid the first egg of her life, at age 10, but it was infertile.

They tried again the next spring and this time she produced a fertile egg, but the chick died just before hatching. In 1979, Tex's egg was soft-shelled and broke. Finally, on May 3, 1981,

Tex laid a fertile egg that hatched into a chick they named Gee Whiz. Tex was killed by raccoons in 1982, but her genes live on: Gee Whiz has fathered many crane chicks, some of which have been released into the wild in the Whooping Crane reintroduction project.

SEE ALSO: *pages 141, 150, 213 and 235 for more on Whooping Cranes.*

Parasitic Parenting and the Mysteries of Imprinting

Q **Last summer I saw a Song Sparrow feeding its baby, only the chick was huge — it looked twice as big as its parent! How common is it for birds to produce such a gigantic baby?**

A What you saw wasn't a giant Song Sparrow — it was a Brown-headed Cowbird. The cowbird female laid her egg in the sparrow nest, and the pair of Song Sparrows raised it as their own chick. When young cowbirds start feeding on their own, they leave their foster parents and join a flock of cowbirds.

Cowbirds are *brood parasites.* They have no instinct to build a nest at all or to care for their babies directly. Rather, the females spend their spring and early summer searching out bird nests in which to lay their eggs. And cowbirds lay a lot of eggs — some may lay as many as 40 per season!

Cowbirds discover nests by sitting quietly, watching the comings and goings of other birds, and also by landing noisily in leaves while flapping their wings, which probably is done to startle and flush nearby birds off their nests. When they find a nest they wait until the brooding female is gone and then rush in. They usually eat or toss out one egg and lay their own. That parent sparrow is stuck raising the cowbird baby.

A Song Sparrow weighs 0.08 ounces (2.4 g) at hatching. It almost doubles its weight on the first day out of the egg, and by day 11, at the time of fledging, weighs two-thirds of an ounce (18.8 g). A cowbird isn't much heavier when it first emerges from the egg, but within seven days weighs almost a full ounce (26 g). It begs more loudly and energetically, and its mouth is larger than the sparrow young, so it elicits more feedings than the sparrow nestlings. Unless food is abundant, the sparrow parents simply cannot provide enough food to meet the demands of the cowbird and all of their own young. Very often at least one of their own chicks will starve.

˅ ˅

Q **Why don't birds just throw out the cowbird egg or hatchling instead of raising it?**

A Most cowbird hosts are smaller than cowbirds, and their beaks are often too small to easily grasp a cowbird egg. In cases where they try, some of their own eggs may get scratched or punctured. A few birds have developed other strategies to get rid of cowbird eggs. When Blue-gray Gnatcatchers detect a cowbird egg, they sometimes abandon their nest and start anew. Yellow Warblers sometimes build a new floor on their nest, relegating the cowbird egg, and any of their own eggs that were there, into the "cellar" where they won't be incubated. Some Yellow Warbler nests have been found with as many as six layers, each with at least one cowbird egg!

But with these strategies, gnatcatchers and Yellow Warblers lose all the eggs they've already laid. Song Sparrows and most other hosts seldom gain anything if they toss out a cowbird egg in an attempt to protect their own young. Researchers have discovered that cowbirds periodically check nests. If a cowbird egg is suddenly missing, the adult cowbird often destroys the remaining eggs or chicks — researchers call this behavior "mafia retaliation." So once a cowbird egg is in a nest, most birds have a better chance of successfully raising at least one of their own if they raise the cowbird.

Another reason that many birds raise cowbirds is that, overall, songbirds have an instinctive urge to incubate eggs and to feed young that they find in their nest. If they were suspicious of any egg or chick that didn't look "right," they might sometimes reject one of their own eggs or young. When most parent songbirds hear the sounds of a begging chick or see a chick's colorful mouth opened wide, they instinctively search for food and try to feed it. In one experiment, European Pied Flycatchers continued to bring food to their own young even after the chicks were satiated when researchers played a recording of begging young. Wildlife rehabbers take advantage of this when they sometimes "foster" a baby bird out by placing it in a wild nest of the same species with young of the same age.

Although many bird behaviors, such as the urge to feed nestlings, are instinctive, these behaviors are refined by learning. The high cost of rearing a cowbird apparently leads some experienced birds to start figuring out that adult cowbirds present a problem. Researchers have found that older Song Sparrow females are more often heard making alarm calls when they detect a cowbird skulking near their nest than younger, inexperienced females do. Ironically, older Song Sparrows are also more likely than young ones to end up with a cowbird egg in their nest, probably because by making those alarm calls, they signal the cowbird that they're nesting nearby.

Q If cowbirds are raised by other birds, why don't they imprint on those species?

A People are still studying this issue. After cowbirds are independent, they join with other cowbirds and don't

The colorful mouths of baby birds elicit feeding behaviors in the parents. In one case, a Northern Cardinal that lost his mate and young spent several days at the edge of a fish pond stuffing food into the colorful mouths of goldfish!

associate with their foster parents any longer. Recent research suggests that female cowbirds may pay attention to their young. There are a handful of records of female cowbirds feeding cowbird nestlings or fledglings, including one record of a banded female feeding a banded chick that was identified as hers. There isn't solid evidence that cowbirds regularly maintain contact with their young, though I've observed adult female cowbirds on the same branch as fledglings and there are many reports of adult female cowbirds feeding fledglings that may have been their own. Whether it's learned or instinctive, cowbirds do apparently know that they're cowbirds and not warblers, sparrows, or other species.

Cowbirds are the only parasitic birds in North America, but worldwide there are several more. Cuckoos, widowbirds, honeyguides, and Black-headed Ducks all must lay their eggs in the nests of other species and leave childcare up to them. Sometimes people use the word "cuckold" in reference to a man who is raising children who are not his own. This word comes straight from a reference to European cuckoos.

Learning about Food

Q Some birds eat all kinds of bugs and seeds, while others ignore what could be perfectly nutritious food and end up starving. How do birds recognize what is and isn't food?

A Some birds are limited in their food choices by specialized adaptations in their bills, feet, ears, and eyes. A hungry Yellow-rumped Warbler may be able to feed on sunflower seed hearts at a feeder, but can't crack open shells. Its digestive system, designed for digesting soft foods, can't grind down the shell if it swallows the seed whole, so yellow-rumps ignore whole sunflower seeds. Unlike other warblers, yellow-rumps can digest bayberries and wax myrtle berries. Some bird bills are designed to extract food items from specific plants, and if birds don't instinctively feed on the right plants, they'll learn to or starve.

Hummingbird bills are an extreme example: their length and curvature can be so splendidly adapted to feeding in a specific flower type that they have difficulties getting food from other flowers. As a result, they have few competitors for the flowers they do use.

Shorebirds have bills specialized for taking different types of food at different depths in sand or mud. Ducks, spoonbills, and some other birds that feed on food items from the bottom of shallow open water have bills specialized to strain out the water, keeping the food items inside the mouth to swallow.

Osprey and Great Blue Herons feed on similarly sized fish but in different areas due to their specialized methods of catching them. When an Osprey catches a fish, it can carry it in its

THE TASTE TEST

Birds that are specialized may be limited in food choices by their own bodies. But even generalists are limited in their food choices. Blue Jays are omnivorous, and during the course of their lifetime travels, they encounter many unfamiliar food items. Some berries and insects that seem perfectly fine may actually be toxic.

To test whether a novel item is edible or not, the Blue Jays I've observed in captivity take a fairly small taste and wait several minutes before eating more. If the item proves to be noxious, the jay will avoid similar items in the future; if not, the jay won't delay in feeding on it the next time it's offered. In one famous case, a captive jay ate a monarch butterfly that quickly made him vomit. After that, the jay always avoided orange butterflies even though other orange species aren't toxic.

talons to the nest. Great Blue Herons have feet adapted for wading and for perching, not for carrying prey. They swallow the fish, fly to the nest, and regurgitate it to their young.

Great Gray Owls normally feed on small mammals, especially meadow voles, which they can hear when the voles are deep in their tunnels, even when those tunnels are buried under 18 inches (46 cm) of snow. The owls' big ears, large but not very strong talons, and huge wings, which can push this fairly lightweight owl out of deep snow, are adaptations for catching their small, hidden prey. When vole populations crash, the owls may "invade" new areas en masse. Some will continue to feed almost exclusively on voles, but a few will learn to take other species. During the winter of 2004–05, hundreds of Great Gray Owls descended on northern Minnesota; some individuals were documented feeding on rabbits, squirrels, and even muskrats, but the majority stayed in grassy areas searching for voles.

v v v v v v v v v v v v v v v v v v v v v v v v v v v

Q How can Bald Eagles survive in northern areas after all the lakes have frozen?

A As much as eagles enjoy fresh fish, they will also dine on carrion and garbage. It may be disconcerting to see the emblem of the United States of America eating at a dump, but the ability of eagles to exploit a wide range of food choices is what makes them so successful.

Crows steal food from other animals. They have been observed distracting a river otter to steal its fish and following Common Mergansers to catch the minnows the ducks were chasing into shallow water. They sometimes follow small songbirds as they arrive from a long migration flight and capture and eat the exhausted birds. They may follow songbirds to their nests to find eggs and young birds. Crows also catch fish on their own, eat from outdoor dog dishes, and take fruit from trees.

EATING ON THE WING

🐦 **The Common Nighthawk** is specialized for taking insects in flight. It has an extremely small beak that is loosely attached to its huge, soft mouth. It opens its mouth wide and flies at moths and other flying insects, which go straight down the throat and esophagus without the bird stopping to swallow. But when the bird is on the ground, the most succulent bug could walk right past without being eaten: the bill is too small and frail, and the vestigial tongue too far back, for even the hungriest nighthawk to pick it up.

When I was a licensed wildlife rehabilitator, I specialized in nighthawk care. To feed these birds when they first came to me, I'd have to very gently tease open the mouth, place mealworms or a special food mash in the back of the mouth, and stroke the throat to help get it down. After a few days, birds would run up to me with their mouths wide open to be fed, but they usually needed several days more to be able to swallow items without help. The wonderful adaptations that allow them to so successfully catch insects on the wing, limit their abilities to eat anything else.

WHO ARE YOU CALLING A BIRDBRAIN?

Here are some examples of ways in which birds are smarter than many people think.

▶ In the wild, jays and crows can recollect when, where, and what food they've stored.

▶ Several species of birds, including nuthatches, have been observed using tools. Brown-headed Nuthatches in the Southeast have been documented using pieces of pine bark to pry off other flakes of bark, to reveal insects beneath. Pygmy Nuthatches in the West have been documented using a twig to probe crevices of a pine in search of insects.

▶ At least one species, the New Caledonian Crow, actually makes tools. One fascinating captive individual named Betty can fashion a hook out of a piece of wire to reach and pull food out of a tube. In the wild, these birds hold pointed sticks or needles in their beaks to extract insects from logs.

▶ Some herons catch fish by setting out bait for them. One popular YouTube video shows a Green Heron dropping bread in the water and moving it about while waiting for fish to nibble at it.

▶ Crows in urban Japan drop hard-shelled nuts onto intersections; they wait for them to be cracked open by cars, and then retrieve them while the cars are stopped after the light turns red.

▶ A few species in the crow family have been documented dropping whelks (large marine snails) and other shelled animals from the air onto rocks and other hard surfaces; the drop breaks the shell and then the crow flies in for a meal.

▶ Western Scrub-Jays can attribute their own motives to other scrub-jays — those jays that take food from other jays' food stores are more likely to keep moving and hiding their own food stores to prevent stealing than are "honest" jays.

▶ Researcher Irene Pepperberg's African Gray Parrot, Alex, could identify by word fifty different objects, could recognize quantities up to six, could distinguish seven colors and five shapes, and understood the concepts of *bigger, smaller, same,* and *different.*

▶ Many pet birds, including parrots, mynahs, and magpies, have learned to imitate human speech. Many owners have long insisted that their birds use words in the proper context. Although authorities have typically pooh-poohed this, Alex's well-documented ability to use language is making many scientists take a second look at human language use in other species.

Birds Do the Strangest Things

Q I saw the funniest video of a bird doing a "moon-walk." Was that real or was it trick photography?

A You were seeing a video of a Red-capped Manakin, a small, plump, colorful bird of Central and South America, doing its courtship dance. The video, which was first aired on an episode of *Nature*, shows animal behaviorist Kimberly Bostwick in the field with three different species of manakins. Each of the species has developed unique and fascinating sound and visual displays, enhanced by specialized feathers and movements, designed to attract females and show off the male's fitness.

Females select as mates the best dancers, which probably increases the likelihood that their young will carry the highest quality genes. Manakins make snaps, whistles, and other interesting sounds and jumps that are too fast for the human eye to follow, so Kim filmed them with high-speed video that captures the action at 500 frames per second, showing that the weird buzzes and clicks are produced not vocally but by the vibration of the birds' wingtips, which can move faster than a humming-bird's wings. The "moonwalking" Red-capped Manakins take a series of quick backward steps to achieve that Michael Jackson effect.

Is the behavior innate or learned? In most manakin species, two unrelated males form a partnership in which they sing and dance in a complex, coordinated pattern unique to their species, like the red-cap's moonwalking. In a manakin partnership, one bird is dominant and gets to mate with the majority of females. The other bird is a sort of apprentice, apparently learning from the dominant male and perfecting his own display.

v v v v v v v v v v v v v v v v v v v v v v v v

Q When I was walking my dog we came upon an injured Killdeer. At least it acted injured. I thought I should bring it to a rehabber, so I followed it, but suddenly it took off and flew away! One of my friends said Killdeer do that when they're nesting. Was it really faking us out on purpose?

A Killdeer and several other species of birds, from Ostriches to songbirds, perform a *distraction display* when potential predators come near their eggs or chicks. Feigning injury with loud calls and drooping wings, the bird hobbles persistently away from the young, drawing a curious person or hopeful carnivore in a direction away from the nest. Killdeer seem to match the speed of the approaching animal, moving more slowly when leading people away than when leading dogs away.

Interestingly, Killdeer seem to modify this behavior for herbivores. A cow or bison is unlikely to eat a Killdeer egg but very well may trample it, and cattle are not likely to follow an injured bird regardless, so when a cow approaches a nesting Killdeer, the bird squawks and attacks it. There is even one report of a Killdeer posting itself directly in front of a nest, emitting a loud, high-pitched squawk, and parting a stampeding herd of bison.

Birds engaging in broken-wing distraction displays perform more intensely as the eggs grow closer to hatching and as the chicks get older; the behavior ebbs as the young learn to fly and are able to escape danger on their own.

✓ ✓

Q My wife is a birder and when we went to Florida, she wanted to see a rare bird, the Florida Scrub-Jay. We went to a park where the birds are supposed to be and quickly spotted one perched at the top of a tree. She was about to take pictures but as she got her camera set up, a noisy group of people arrived. Instead of flying away, the bird flew right up to them and suddenly in flew five more scrub-jays! They actually landed on the hands of all these noisy people, taking peanuts that the people had brought to offer the birds! Are these birds endangered because they're too friendly or curious for their own good?

A Florida Scrub-Jays are threatened because their habitat, sandy areas of Florida covered with native scrub vegetation such as palmettos and evergreen oaks, is being destroyed for citrus groves, housing projects, and shopping malls. These jays are adapted physically and behaviorally to this special habitat, which once covered much of central Florida and was maintained in low, open condition by frequent wildfire. Now at least 35 of the plant species in their scrub habitat are listed as endangered or threatened.

These jays are highly sedentary — the most successful ones never move far from their parents' territory, staying within a total area of about one square mile their entire lives. When their territory is developed, the jays must disperse but cannot find a new territory because all the scrub around them is already occupied by territorial jays.

As human housing projects multiply, the remaining scrub grows tall and dense because of fire suppression management. These areas become unsuitable for the jays, and predators such as house cats have an easier time catching the young birds, so the local Florida Scrub-Jay population becomes extinct. This is happening in every region where this species exists.

Florida Scrub-Jays feed on a variety of arthropods, small vertebrates, berries, and acorns. Their jaw support and beak shape are specially adapted for opening acorns, which they can feed on year-round, even when acorns are not in season. During the fall acorn season, each jay stores thousands of acorns in the sand all around its territory. During the winter, when insects are scarce, they dig up and eat their stored acorns. They're also known to land on the backs of deer, cattle, and feral hogs, to pull off ticks to eat.

They also frequently land on people who offer acorns, peanuts, or other food. The bird you and your wife found first was a sentinel, watching for predators and rival jay families. When the people arrived with peanuts, the bird probably alerted its family and they quickly appeared in hopes of a handout.

∨ ∨

Q **I spent a day at an ocean beach watching as Brown Pelicans dived straight into the water, dozens and dozens of times. I've been watching American White Pelicans for years without ever seeing them do that. Why do two birds that look so similar act so differently?**

A American White Pelicans are specialized for feeding in freshwater, usually in shallow areas of lakes and rivers. If they tried diving into those waters, they'd kill themselves! Instead, they often group into little squadrons and fish cooperatively, forming a tight line and, by beating and dipping their wings and bills, herd fish into the shallowest waters where they can scoop them up.

Brown Pelicans, on the other hand, are marine birds, and although they, too, are extremely sociable and nest and loaf in large colonies, they don't feed cooperatively. Instead, this species has perfected plunge diving. Brown Pelicans can see fish beneath the surface and can calculate where to dive, correcting for the refraction of light that makes fish appear in a different place than where they really are. During the dive, they pull their head in over their shoulders, pull their legs forward, and bend their wings at the wrist. Interestingly, they also rotate their body to the left, which probably protects their trachea

and esophagus as they hit the water. These vital structures are situated on the right side of their neck.

As their bill enters the water, the birds thrust their legs and wings backward, moving their bill toward the fish even faster. Their huge throat pouch expands as it fills with up to 2½ gallons (9.5 l) of water; the pressure forces the lower bill into a distorted bow shape but it doesn't break, thanks to special muscle adaptations.

The streamlined upper mandible guides the fish into the mouth, the lower bill bounces back into shape, and the bird closes its bill, trapping the fish inside the pouch. It raises the back of its head slowly with the pouch pressed against its breast to drain out the water while retaining the fish, then tosses its head up to swallow the fish. If the dive was unsuccessful, the bird quickly raises its head with the bill open so the water drains out immediately. It takes less than 20 seconds to drain the pouch and swallow the fish.

❯ ❯

Q Do birds play?

A Many mammals engage in "play," that is, activities that enhance learning of motor and sensory skills and social behaviors but otherwise serve no immediate purpose. Young screech-owls pounce at leaves; young crows and jays pick up, inspect, and hide all kinds of shiny objects; young gulls and terns carry small items aloft and drop them, catch them in midair, and drop and catch them again. All these activities probably help birds acquire the skills and coordination they'll need for hunting and other essential activities as adults.

Some forms of play, called "locomotor play," seem quite similar to the exhilarating play of children sledding down a steep hill. Some ducks have been observed floating through tidal rapids or fast-moving sections of rivers, and when they've reached the end, hurrying back to the beginning to ride over and over. In the air, ravens and crows often rise on air currents only to swoop down toward earth, then glide back upward, again and again.

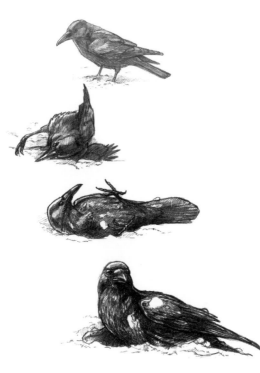

Common Ravens have been observed taking turns sliding down a snowbank on their tails or rolling over and over down a hill.

LEARNED VS. INSTINCTIVE BEHAVIOR

I receive many questions about learned vs. instinctive behavior. By "instinct," people usually mean innate behaviors — those a bird does in a particular situation without learning or trying that behavior beforehand. Many innate behavior patterns are refined by learning.

For example, when a robin first hatches, it already knows how to do three things.

▶ If something lands with a soft thud or jostles the nest, the little bird pops up like a jack-in-the-box with its mouth wide open to be fed.

▶ As soon as the little bird swallows, it backs up and poops.

▶ After it poops, it crouches down again and remains fairly motionless until the next time the nest is disturbed.

On a visit to the nest, the parent robins feed whichever nestling begs first, extends its neck highest, or holds its beak closest to the parent's beak. Adults usually follow one or two flight patterns, alighting in the same spots over and over, so nestlings quickly learn, by feeding success, where to direct their beaks. They're already refining their innate behavior by learning.

Learning Who's Who and What's What

For several days after hatching, robin nestlings beg if anything, including people or predators, alights on the nest or overshadows them. About five days after hatching, their eyes open and they start noticing their parents and one another. Once they recognize the birds that feed them, they will crouch down if anything else shows up. By the time they're 10 or 11 days old, if anything approaches the nest they'll fly off in a flurry, though they can't go far and can't get back up to the nest. (This is why it's so important to not peek into robin nests after the young are about a week old.) If the young survive this sudden, panicky escape, their parents will continue to feed them and will try to lead them to dense vegetation to keep them hidden.

Robins don't "imprint" on their parents or each other. For a day or two after leaving the nest, especially if something caused them to leave the nest too soon, a hungry robin chick who cannot find its parents may beg from other bird species or even people. I once saw a robin fledgling begging from my golden retriever! After a robin is flying well, it learns to follow its parents and associate with other robins. After that, it avoids other species.

Robins have a few sharp warning calls, including a high-pitched *seeee* made when a hawk flies overhead, and *peek* and *tut-tut* calls when danger is lower to the ground.

(continued)

Young birds recognize the sounds their parents make and act on these sounds; some of this may be innate and some refined by learning as they notice what their parents do.

Both parents attend to the young after they leave the nest. After the female produces a new clutch, she incubates them as the father stays with the fledglings. Every night he flies to a communal robin roost, and they follow. He will start focusing on his new nestlings after they hatch; by then the fledglings will be independent.

Scratching and Flying

As its feathers start to grow, a young robin scratches with its beak and claws, an innate behavior that is clumsy at first. The bird quickly improves its technique by learning what movements are most comfortable and effective. The young bird innately flaps its wings, and the combination of feather growth, developing muscles, and learning refines those initial awkward efforts into skillful flying.

Making More Robins

Robins are drawn to other robins in big, loose flocks. They stay safe and discover new food sources thanks to the many eyes and ears of the group, until rising hormone levels in spring induce territorial and breeding behaviors. These behaviors are innate but are also refined by the responses the robin gets and how successful each behavior is. Singing robins refine their songs and add phrases unique to the neighborhood they nest in, even though they may have been raised far from there.

When robins lose their young, they move to a new site to re-nest, and sometimes a pair breaks up. If they successfully raise a brood, the same pair will often re-nest two or even three times that season. Although robin pairs work together in raising their young, their bond isn't quite what romanticists might like. In one study, "extra-pair paternity" occurred in almost 72 percent of all broods. In each of these "EPP" nests, at least one chick wasn't fathered by the male taking care of it. According to that study, females may be allocating paternity based on their assessment of each male's parental performance. Females invest a great deal into their eggs, and multiple fathers hedge their bets.

If a mate is killed, sometimes another male will take over rearing the chicks. In my own backyard, a Cooper's Hawk once killed the male robin nesting in my hedge. The next day, the female had accepted a new mate who helped her feed her nestlings. I wondered if he was the genetic father of one or more of the nestlings, but he apparently never insisted on a paternity test.

Species Specific

Q I am curious about the Turkey Vultures that roost in the tall pine trees a block from our house, right near a cemetery — there is even a sign on the road that says "Dead End"! I never thought of vultures as particularly social — what are their mating and nesting habits?

A Turkey Vultures usually forage individually. By splitting up, each bird sniffing out carcasses, one is eventually going to hit the jackpot and others will notice it dropping down and will join it. In the evening, vultures join together at communal roosts, ranging in size from a few birds to several thousand. Spectacular evening flights "advertise" these roosts.

One 15-year study of tagged Turkey Vultures in Wisconsin suggests that they usually mate for life, but when a bird's mate dies, it does take a new mate. Although they spend the nesting period together and cooperate closely in incubating eggs and caring for the young, there's no evidence that a pair migrates or spends winter together.

Turkey Vultures nest on cliffs, hollows beneath fallen logs, rock outcrops, caves, abandoned buildings, and other places isolated from humans. They don't construct much of anything in the way of a nest for their (usually) two eggs. Both parents develop brood patches, a bare spot on their underside that is pressed against the eggs. Both take turns incubating the eggs, which take over a month (often as long as 40 days) to hatch.

The parents feed their fluffy white chicks the same rotten meat they eat, only in regurgitated form. The chicks make their

AS SILENT AS THE GRAVE

Turkey Vultures lack a syrinx and the voice-producing muscles associated with it, so they are essentially always silent. They can make a guttural hiss when agitated, either by a disturbance at the nest or when two birds are vying for the same part of a carcass. And nestlings make a characteristic hiss that is almost inaudible while they're still blind and unable to hold up their heads. By the time they're a week old or so, they make a specific nestling-hiss when disturbed. This sound has been variously described as a persistent and vigorous wheezing-snoring; a low, throaty, or growling hiss; a snake rattle; or a roaring wind. The sound volume and quality depend on the age of the nestling, the intensity of the hiss, how close the observer is, and the acoustic qualities of the nest area. The sound is often given by both nestlings at once. If disturbed, nestlings can also stomp their feet or loudly flap their wings.

first flights sometime between 60 and 80 days after hatching. After they leave the nest, they receive little or no care from their parents, but siblings sometimes stay together for a while. They join communal roosts, keep track of other birds in order to locate food as they get practice sniffing out the world, but are generally on their own.

Scientists are still trying to tease out the relationship of American vultures to hawks and to storks. Traditionally, vultures have been classified with hawks, based on many features and their need for meat. But in the 1990s, they were placed in the same order as storks based on many similar features, such as bald heads, perforated nostrils, and a curious habit of urinating on their legs to cool off. Then thanks to reanalysis of DNA, they were placed closer to hawks again.

Regardless of where they really belong, it's appealing to consider the similarities of storks, the birds that most symbolize birth, with vultures, the birds that most symbolize death. And it's also interesting to note that vultures, as well as crows and ravens, seem to avoid scavenging the carcasses of other large black birds.

˅ ˅

Q I often see a crowd of Cedar Waxwings and/or robins on my crabapple trees in the late winter, but once I saw two pairs of Pine Grosbeaks. They stayed for about 20 minutes, eating the fruit and bathing in an icy puddle in the yard. What is their usual habitat and might I see them again here in western Massachusetts?

A Pine Grosbeaks are members of the finch family, along with goldfinches, redpolls, siskins, Purple and House finches, crossbills, and Evening Grosbeaks. They breed in subarctic and boreal forests in Asia, Europe, and, in North America, from eastern Canada to western Alaska. They also breed in coniferous forests of western mountain ranges and in coastal and island rain forests of Alaska and British Columbia. Like other northern finches, Pine Grosbeak migrations are unpredictable. But they move southward less often than, and don't go as far south as, many other "winter finches."

They aren't particularly drawn to robins or waxwings but are attracted to mountain ash and crabapple, and so are often found feeding with them in winter. Male and female Pine Grosbeaks are very territorial during the breeding season — both will fight off members of their own sex that invade their territory. But after the young fledge, Pine Grosbeaks become gregarious and associate in flocks. It's possible that their winter

A MAGIC MOMENT

The very first Pine Grosbeak I ever saw was a young male separated from his flock. I heard him before I saw him and whistled back to him when I was still more than a block away. I walked toward the sound of his whistle, continuing to answer his persistent whistles with as good an imitation as I could muster. Soon I saw him at the top of a tree, looking right at me when I whistled. I have no idea what prompted me on a frozen February day, but I took off my glove and raised my hand, and he lighted right on my finger! He looked at me and whistled, I whistled back, and he gave me one last whistle before flitting back into the trees.

It was one of the most thrilling moments of my life. But it may well have been one of the most disappointing for the grosbeak. I think my whistling back to him must have given him hope that he'd found another grosbeak flock, but all he found was a clumsy, flightless human.

flocks are composed of family members. As I noted, they come south occasionally but not predictably. When they appear in Massachusetts, it's something to treasure.

˅ ˅

Q Why are Saw-whet Owls so small?

A Bird size is determined by a combination of things. One of our largest owls is the Great Gray Owl, which specializes on pretty much the same size food that Northern Saw-whet Owls eat — very small rodents. The Great Gray Owl takes advantage of its huge ears to hear meadow voles when they are buried under deep snow or meadow grass; it flies in and plunges to grab the vole, then uses its huge wings to pull back out. The problem with having such a large body is that it must catch a lot of mice to survive.

Northern Saw-whet Owls are too tiny to hunt where snow or grass is so deep, but they live in forests where snow isn't as deep, especially around tree trunks, so although they don't eat as many mice as Great Gray Owls do, they can find enough to keep their tiny bodies alive.

Their size makes them very maneuverable in flight, and they can also obtain a good amount of the nutrition they require from large insects in summer. In addition, Northern Saw-whet Owls are small enough to take shelter in woodpecker holes, which Great Gray Owls certainly can't do!

Bird feathers insulate the bird, protecting it from too much heat and cold, but they can also hold an incubating bird's heat away from its eggs. So at the time eggs are laid, belly feathers on most female birds, and the males in some species, become loose. The parents pluck them out, often to use as part of their nest lining, so they can press their warm abdominal skin directly on the eggs to heat them.

Q I live 150 miles from the coast. Why do we have seagulls flying around a local strip mall parking lot? I see them all the time.

A The term *seagull* is technically correct only when referring to Jonathan Livingston or to a gull that happens to be on the ocean, but it's widely used as a generic term for gulls in general. A great many gulls spend their lives inland. The one that usually frequents fast food restaurants and strip mall parking lots is the Ring-billed Gull. It feeds on slugs, worms, and other invertebrates found on lawns, scavenges on garbage, and mooches from people. I've watched Ring-billed Gulls pluck warblers out of the air first thing in the morning when the exhausted, tiny songbirds are coming down after a night's migration. They also feed on fish. Their mouths are very wide, and their throat and esophagus expand to swallow an entire small fish.

CHAPTER 6

Singin' in the Rain: Bird Vocalizations

On a frigid winter morning as I walk through the woods, when even the wind is silent, the sound that penetrates my warmest fur-lined hat is the rich song of a cardinal, or another song lovely in its simplicity — the *Hey, sweetie!* of a chickadee. By spring, birdsong erupts well before dawn, and the morning chorus has so many voices that it may be hard to pick out that same cardinal and chickadee, but they're in there. The beautiful voices that fill the spring air make the whole world seem alive and happy.

How do birds produce such complex sounds? Why do they sing? And how do we learn to identify their voices? Those are the questions people send me.

^v^v^v^v^v^v^v^v^v^v^v^v^v^v^v^v^v^v^v^v^v^v^v^v^v^v^v^v^v^

Making Beautiful Music

Q I went on a bird walk and my leader pointed out a Wood Thrush song. It was cosmically beautiful! How can a single bird produce such an amazing sound?

A Because birds produce their sounds with their syrinx, or song box, which has two branches and two sets of muscles, they can produce harmony with their own voice. Birds in the thrush family have very complex muscles in their syrinx, allowing them to produce breathtakingly complex tones independently in each branch.

That morphological description may begin to explain how the individual tones are produced, but what can explain the sheer beauty of these sounds? Do male Wood Thrushes take as much pleasure in singing as we do in listening to them? Do female Wood Thrushes select mates for a dry, analytical reason such as how many different tones in what arrangement each male sings, or because the beauty of a song takes her breath away? Science gives us a lot of answers, but it might take the Vulcan Mind Meld to resolve that one.

⌄ ⌄

Q I live in the woods in northern Minnesota, and in May and June I usually wake up to a Winter Wren singing near my cabin. How does such a tiny bird produce so many sounds so quickly?

A As with other splendid bird songs, our experience of Winter Wrens transcends a mechanical understanding of sound production. In 1884, the Reverend J. H. Langille described his experience listening to the Winter Wren, "I stand entranced and amazed, my very soul vibrating to this gushing melody, which seems at once expressive of the wildest joy and the tenderest sadness."

Per unit weight, Winter Wrens have ten times the sound power of a crowing rooster, and birds in the Eastern population sing a good 16 notes per second — an impressive output that is not only exceeded but more than doubled by Western birds, which sing 36 notes per second! Their rapid heartbeat, respiratory rate, and metabolic rate don't explain the output, since larger and smaller birds don't match this! But what is even more amazing is that these birds not only produce the sounds but react to tiny parts of the songs, so their ears and brain can resolve in real time individual notes that we cannot without replaying the songs at slow speed.

ˇ ˇ

Q Why are bird calls so varied? Cardinal and Mourning Dove calls are so different from each other, for example. Is there something in their physical makeup that makes the sounds distinct?

A Yes. The size of the bird and length of the trachea and bronchial tubes impact the sound quality and frequency, and produce sounds in their syrinx, which is far more complex than our human larynx. Although our larynx is a simple instrument, made of a muscle that spans the trachea and can be controlled to make varying sounds, human voices are still so variable that most of us can recognize dozens, and sometimes hundreds, of individual people by their voice.

The syrinx is located where the trachea branches into the bronchial tubes and, depending on the species, may have a variety of muscles arranged within this more complicated framework. When a group of people sing together, we make up a chorus. When birds do, it's more like a whole symphony orchestra.

ˇ ˇ

NAME THAT TUNE

🐦 **Describing bird sounds can be tricky.** For example, I once received this question: "Could you tell me what bird is most often the first to sing in the morning, and sings all day. I hear it most of the year except in winter."

Robins are usually among the very first birds to pipe in with the dawn chorus, so I sent him to a Web page with the robin song and told him that if this wasn't it, to please try to describe how the sound of his bird was different. He wrote back, "This bird has a two-sound cadence — a short sound and a very long ending that runs for maybe two seconds. I normally hear this bird near water or marshes." I still didn't have a sense of its tonal quality, so I asked if it sounded buzzy like a Savannah Sparrow, with another link. He responded, "No, it's more melodic — a long pitch sound like the bird is saying *hel-lowwwwwwwwwwwwwwww.*"

Then it hit me. He was describing a White-throated Sparrow. When I sent him a link to that sound, he was ecstatic. Detective work can be ever so rewarding.

Scientists describe bird songs using sound spectrogram, or a *sonogram* — a graph showing each tone's frequency and duration. Some practice or training are necessary to interpret them.

Rather than describe a song at all, people now often email me recordings. They don't need a fancy microphone — many just point a digital camera toward a singing bird and use the video function. Whether or not we can see the bird, this lets us hear it!

Q My wife and I moved to Dallas in January. As the weather has warmed, we've come to expect a high-pitched *twirp* in the twilight at dusk and dawn. It sounds like some swallow, and I assume that it's feeding on insects, though I don't know how it sees them in the darkness. I don't think this is bats — there are too many. How can I figure out what bird this is?

A It's very tricky to describe sounds in a way that someone else can "hear." But I would bet your birds are either Common Nighthawks or Chimney Swifts. Nighthawks make a funny *beep!* or *peent!* in the twilight sky as they catch insects. Chimney Swifts also feed in low light, and you're apt to hear them because they gather in large congregations to roost. How can you tell which one this is unless you get a good look? Go to *www.allaboutbirds.org* and type in just about any North American species name and you can listen to its sounds.

> The number of different songs that a bird sings varies depending on the species and the individual bird. Chipping Sparrows repeat the same single song over and over. Northern Cardinals sing 8 to 10 songs, American Robins 70, and Northern Mockingbirds 200. Brown Thrashers are the record-breakers, however; they are mentioned in *Ripley's Believe It or Not!* for their amazingly huge repertoire of 2,000 songs.

FEMALES WHO SING

When I took an ornithology class back in the 1970s, we learned that male birds do the singing and females do the listening. This seems obvious intuitively: males are the ones with conspicuous plumage and the job of defending the territory while females incubate eggs, usually trying to be as inconspicuous as possible. And it's true that many or even most songs we hear from familiar backyard birds, from mockingbirds to Chipping Sparrows, are sung by males.

But females of many tropical species sing, and also females of several northern species, including Northern Cardinals, Black-headed Grosbeaks, Rose-breasted Grosbeaks, and White-throated Sparrows. Scientists are still teasing out the reasons why some females sing while others don't — this is yet another area where the more we learn, the more questions we come up with.

Female White-throated Sparrows have a very interesting and unusual singing pattern. The species has two color forms — one has white stripes on the head and the other has tan stripes. This is a genetic difference like that underlying human eye color. Birds of either sex can have head stripes of either color.

Both male and female sparrows with white head stripes are more aggressive than tan-striped birds. Female white-striped birds sing, but tan-striped females do not. In laboratory tests where birds were separated by a glass partition but could see those of the opposite sex through the glass, females of both color forms preferred tan-striped males, but the more aggressive white-striped females outcompete the tan-striped females for tan-striped mates. Those same laboratory tests indicate that males prefer the appearance of white-striped females, but that when white-striped females start to sing, the aggressive white-striped males attack them. So virtually all pairs of White-throated Sparrows have a male and female of opposite color forms.

What Do Birds Sing About, Anyway?

Q The cardinal outside my window has been singing all morning. Why does he spend so much time singing?

A He sings to attract a mate and to announce his presence to other males so they will stay off his territory. Some birds sing tens of thousands of songs each day. Defending a territory requires constant vigilance!

If you listen carefully, you may hear *countersinging*, when neighboring birds sing in response to each other. When Marsh Wrens countersing, they may match each other by singing the same song type as a neighbor. Or they may sing an entirely different song type, or anticipate the song the other will sing next and sing it first. All these choices allow males to communicate their intentions before deciding whether to fight. Depending on how they sing, they can direct threats to specific neighbors or back down from a challenge.

In addition to vocal sparring, birds may use their repertoires to impress potential mates. Male Song Sparrows with larger repertoires have a longer lifespan than do males with fewer songs. This is probably because males that are better nourished and defend better territories are able to learn more songs.

Young male Bewick's Wrens that hatch early in a season have more time to learn songs before winter than later-hatching males do; those with more songs in their repertoire can successfully defend the best territories. In these species, females prefer to mate with the males with the most song types, an easy way of discerning which males are robust and experienced, and thus, most likely to help her successfully raise healthy young.

˅ ˅

Q **My son says he's been tired lately because the birds wake him up at dawn. Is it my imagination, or do the birds sing especially loudly at first light?**

A It's true: more birds sing at dawn than later in the day, and they sing with more energy and variety. This "dawn chorus" actually starts an hour or so before dawn in spring and early summer. The chorus often begins with American Robins singing a much more rapid, excited version of their daytime song. Chipping Sparrows sing their songs at a frenetic pace, and many other birds sing with exceptional energy as well. As more birds join in, the chorus crescendos. At its peak, the dawn chorus can be so richly complex that it can be tricky for human ears to pick out many of the individual voices.

Ornithologists are still debating why birds sing so vigorously during the dawn chorus. Do birds simply have a lot of pent-up energy after a good night's sleep? Is dawn the best time to sing because in the dim light, territorial competitors and prospective mates don't have much else to do other than listen? Is morning the best time to make a statement to competitors and potential mates that may have landed in the morning after a night of migration? These exceptional dawn choruses are also most common in the temperate zone, where birds have a compressed breeding season lasting only a few short weeks.

Whatever the reasons for it, the dawn chorus is one of the truly spectacular yet everyday occurrences that most of us take for granted. If the birds wake you up in the morning, consider stepping outside to listen, or even getting up before dawn and going to a natural area near you, where you'll experience an even more unforgettable show.

More than Just Melody

Q I've heard mockingbirds singing songs that I know belong to other birds — what's going on? Why do they imitate so many other sounds?

A Although some birds learn their species' song during their first year of life, others, including mockingbirds, continue adding to their repertoire as they grow older. Northern Mockingbirds can learn as many as 200 songs and often mimic sounds in their environment including other birds, car alarms, and creaky gates. One theory is that if a female prefers males who sing more songs, a male can top his rivals by quickly adding to his collection some of the sounds around him. Possessing a diverse assortment of songs may indicate he is an older male with proven longevity and survival skills — good traits to pass on to offspring. An older male may also be more experienced in raising young or may have access to better resources.

According to one study on the Edwards Plateau in Texas, mockingbirds with the largest selection of songs have the best territories, laden with foods such as insects, wild grapes, and persimmons. One 2009 study found that mockingbird species in areas with unpredictable or harsh climates include more imitations in their songs than those in milder climates. In harsher areas, it may be more critical for females to be extremely choosy about which mates they accept.

Some researchers have suggested that mockingbirds may use other species' songs to warn those species to keep away from their territories, but this possibility has never been thoroughly investigated. There is no evidence that a mockingbird song has ever caused a cell phone to head to another territory, even though they imitate those!

At least four American Presidents had mockingbirds as pets in the White House: Thomas Jefferson, Rutherford B. Hayes, Grover Cleveland, and Calvin Coolidge.

191

MORE MIMICS

The Northern Mockingbird is the best-known mimic in North America, but starlings, in the same family as the mynah, are also famous for their mimicry. Wolfgang Amadeus Mozart had a pet starling that could mimic tunes and make variations of them. Renowned mimics, such as the lyrebirds of Australia and the Lawrence's Thrush of South America, occur on other continents, too.

Male Marsh Warblers learn the sounds of other species on their wintering grounds in Africa. Perhaps these varied sounds impress potential mates when they return to breed in Europe. Indigobirds in Africa are also mimics, but for an entirely different reason. Indigobirds are brood parasites that lay their eggs in the nests of other species. For example, the Village Indigobird lays its eggs in the nest of the Red-billed Firefinch. Young indigobirds learn the begging calls of the firefinches that raise them so they will not be recognized as an intruder. Young male indigobirds also mimic their hosts.

The female Thick-billed Euphonia is a Neotropical bird that imitates the alarm calls of other species when her nest is threatened. These sounds may get the attention of other species to help in the attack of a predator or other perceived threat.

Some species not typically thought of as mimics sometimes imitate the vocalizations of other species. Blue Jays imitate the calls of Red-tailed, Red-shouldered, and Broad-winged Hawks, for example. The function of these imitations is unknown, but sometimes jays call out a raptor imitation just before flying into a feeder, and when they do this, the birds at the feeder often scatter.

Some observers have noticed that when jays imitate hawks, incubating birds sometimes fly up from their nests, so this mimicry may help jays to discover where nests are. Jays raid nests for eggs and nestlings during their own breeding season, when protein is critical for their own young.

In some cases, mimicry may result from the song-learning process gone awry, such as reports of a Vesper Sparrow and House Wrens singing songs of the Bewick's Wren, and an Indigo Bunting and a Common Yellowthroat singing a Chestnut-sided Warbler song. It seems that a fairly large number of these occasional mimics are unpaired, suggesting that males who learn the wrong songs often fail to pass their genes to the next generation. Selection against birds who learn the wrong songs may thus be very strong, so "mistakes" are not perpetuated.

Q When I went on a birding trip to the tropics, our guide told us to listen to the sounds of birds that he said were a male and a female wren duetting. Do our backyard birds duet too?

A Interestingly, duets are far more common, complex, and coordinated among tropical birds than they are among species that breed in the temperate zone. For example, a female Red-winged Blackbird may join in a duet by uttering a loud series of notes during the last half of her mate's *oakalee* song. In contrast, some tropical wrens sing such highly coordinated duets that unless the birds are far apart, it's hard for a listener to even realize that more than one bird is singing. Other tropical duetting species include some parrots, woodpeckers, antbirds, flycatchers, shrikes, and wrens.

Birds that are resident in the tropics are much more likely to maintain long, monogamous bonds and to live year-round on their territory. This may provide more opportunity for them to develop complex duetting, which may further strengthen their bond or aid in joint defense of the territory.

˅ ˅

Q I've often heard the Song Sparrows behind my house singing a beautiful melody, but I was surprised when I heard one make a short little chirp that sounded nothing like the songs. Was it a young bird that didn't know how to sing?

A	Not necessarily. You heard a contact call, a sound that the birds use to tell one another, "Hey, I'm here." Males, females, adults, and juvenile birds may all use this call year-round. In contrast, it's typically only male Song Sparrows who sing the more complex melodies. They only sing during the breeding season, to advertise their presence to females and to warn males to stay off their territory. (In rare cases, female Song Sparrows sing, but their song is usually simpler than that of the males.)

Sparrows may also use a variety of other vocalizations to communicate with one another. For example, Song Sparrows utter a high-pitched alarm note that ornithologists describe as a *tik* when a hawk flies overhead. They use a lower-pitched call described as *tchunk* when humans approach a nest or fledglings. Females make a harsh chattering sound to their mates during nest building, and they utter a nasal trill after mating. Both males and females "growl" as a threat. When migrating at night, Song Sparrows also make high-pitched *tseep* calls, perhaps to alert nearby birds to their presence in the dark skies.

Paired-up goldfinches make virtually identical flight calls; goldfinches may be able to distinguish members of various pairs by these calls.

SINGING ON THE WING

Some birds, especially those that live in expansive grasslands or tundra with few conspicuous perches, sing in flight. This allows their voice to carry farther and provides them with a visual as well as auditory display.

At least one forest bird, the Ovenbird, a warbler living in forests in eastern North America, also has a flight song, usually given at dawn, dusk, or nighttime. The bird starts out singing anywhere from ground to mid- or even upper-canopy and suddenly takes off in labored flight, wings and tail outspread as he circles and continues to sing. This display has been nicknamed the Ovenbird's "ecstasy flight." The flight song is quite a bit different from the normal *tea-CHER, tea-CHER, tea-CHER* song produced most of the day. The function of this song isn't understood yet.

Many male hummingbirds have aerial displays, combining a specific flight pattern, often a deep U, with a wing buzz. Male Common Nighthawks display by flying toward the ground and suddenly making a loud boom, produced by air rushing through the primary flight feathers after suddenly flexing the wings downward during the dive.

Male American Woodcocks may engage in evening "skydances" during the spring. Several males gather in an open area near the woods where they spent the day.

When the light grows very low, they start producing a buzzy *peent* vocalization from the ground. Then, one bird will take off, his wings making a lovely chittering sound as he spirals toward the sky. Suddenly he breaks into a chirpy vocalization and drops to the ground to begin anew. Female woodcocks are drawn in by this display.

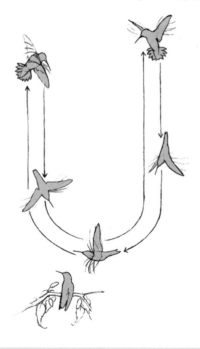

Q I heard a tremendous racket from birds up in a tree last spring. When I got up close, I could see that they were fluttering around an owl. Why do they make so much noise? Shouldn't they try to keep quiet and stay away from the owl?

A The birds you heard were uttering alarm calls after they noticed the owl. Many songbirds recognize alarm calls not just of their own species, but of many other birds. They may join a mob of birds making harsh scolding sounds while dive-bombing an owl, snake, cat, or other predator. Together, they may drive the predator away, or at least distract or harass it so that it won't discover their nests or young nearby, or where they themselves go to roost at night.

Q Once I was walking in the woods and heard a soft, low thumping sound that got faster as I listened. I couldn't see what was making the sound, but a friend told me it was a Ruffed Grouse. Is that true?

A Yes, male Ruffed Grouse produce a distinctive drumming sound, beginning slowly at first but speeding up into a whir. Because grouse often stand on a log while doing this, people once believed this sound was produced by the grouse's wingtips beating the log — or that the bird thumped its wings against its breast, Tarzan-like.

But in 1932, Arthur A. Allen hid out near a grouse's drumming log and caught a grouse on film with his slow-motion movie camera. The stopped frames of the film showed that the grouse stood crosswise on the log, braced on his tail, and cupped his wings, bringing them forward and upward with such force that he compressed a parcel of air between his chest and wings, creating the sound wave without touching his wings to his chest or to the log. The sound is deep, about 40 hertz, which is at the low range of human hearing, explaining why some people say they almost "feel" the sound as much as hear it.

In spring, Ruffed Grouse males declare their territories and announce their presence to females by drumming. Although drumming peaks in April or May, they may drum at any time of year. In fall, you can often recognize a young grouse by his pitiful attempts at drumming. Until a bird masters the trick of cupping the air properly, he may simply produce a dull flapping sound, but as with many skills, practice makes perfect.

If the idea of producing such a sound simply by compressing air in the "wingpits" seems hard to imagine, remember that humans can also produce a sound in our armpits by compressing air. There's one notable difference between Ruffed Grouse and humans, though — female grouse find their partners' "pit" sounds attractive.

v v v v v v v v v v v v v v v v v v v v v v v v v v v v

Q **I saw the most bizarre thing on my way to work. A woodpecker flew to a stop sign and pounded on it with its bill. Why was it doing that?**

A Woodpeckers usually drum on trees to announce their territories or to attract mates, but the woodpecker you saw took advantage of the metal sign to help broadcast his message more loudly. The farther a woodpecker's drumming sound carries, the larger the territory he can defend. Different woodpecker species drum with distinctive rhythms. For example, Yellow-bellied Sapsuckers begin with a drum roll, pause, then add some slower strikes. After a sapsucker drums, another sapsucker may drum back in reply.

Pileated Woodpeckers drum rapidly, about 15 beats per second, speeding up near the end of the burst. They also communicate using other rhythms, such as with a double tap near a promising nest site, or a series of taps from within their nest cavity as a mate approaches to take a turn at the nest.

v v v v v v v v v v v v v v v v v v v v v v v v v v v v

Q **What's the best way to learn how to recognize different birds' songs?**

A The best way is to track down every natural sound you hear and look for the animal that's making it. When I was a beginner, I spent three evenings trying to track down a beautiful whistle that I *knew* had to come from a bird, only to finally find a one-inch tree frog called a spring peeper!

When you search high and low for ten minutes to find one particular singer, you can't help but remember that sound, bird or not. By tracking down each singer, little by little you'll also absorb clues about the sound, habitat, time of day, time of year, and other factors experts use to identify birds by ear.

You can listen to audio guides or birdsong recordings on the Internet to help you identify the songs of birds in your area or at another birding destination. It's often most effective to listen to a few songs at a time and become familiar with them before moving on to the next few. You can also watch a DVD or CD-ROM with video footage of birds singing to help you associate the bird with its song.

Whether listening to recordings or to birds in their natural settings, you can use mnemonics to help you remember the sound. A few examples include the notes of a Black-capped Chickadee (*chickadee-dee-dee*), the song of a Carolina Wren (*teakettle, teakettle, teakettle*), and the hoot of a Barred Owl (*Who cooks for you? Who cooks for you-all?*).

Many songs don't have a consistent rhythm pattern but are easily recognized by the tonal quality. Crows *caw* while ravens *croak*. Rose-breasted and Black-headed Grosbeaks have a song somewhat similar to an American Robin's, but the grosbeaks have a richer tonal quality than the robin. I describe the difference as that between an opera singer such as Beverly Sills and a movie singer such as Julie Andrews.

Many birds sing a simple staccato trill. When you master the Chipping Sparrow's song, you can compare similar trills as being shorter (Dark-eyed Junco), slower (Swamp Sparrow), more musical (Pine Warbler), more mechanical (Worm-eating Warbler), and so on. Listening to sound recordings of wild birds will help you recognize those kinds of differences.

∨ ∨

Q How do birds learn their songs?

A There are almost as many answers to this as there are species of birds! Some scientists have devoted their careers to figuring this out, species by species.

Some bird songs are hardwired into a bird's brain. Virtually all flycatchers, including kingbirds and phoebes, fall into this category. If Eastern Phoebes are raised in captivity and never exposed to their species' songs, they still develop normal phoebe songs.

Marsh Wrens imitate elements of songs they hear. When a Marsh Wren egg was hatched and reared by House Wrens, it produced unrecognizable songs. When exposed to a training tape, captive Marsh Wrens imitate song elements of a few other species. Wild males engage in "matched countersinging" with neighbors, so they continue to learn song elements as adults.

Sedge Wrens seem to improvise their own unique song repertoires. American Robins share some whistles with neighbors, so they apparently learn song elements from one another. Baltimore Orioles may learn their songs from their fathers and neighboring orioles during their first summer. It's possible to detect differences between songs of yearlings and older males, so orioles adjust their songs at least until they're two years old.

As beautiful as Hermit Thrush songs are, there has been little study about how they are learned. Wood Thrush songs have three parts. The middle part is apparently learned by listening to other Wood Thrushes. The first and last parts are either innate or invented by the birds. When Wood Thrushes are hand-reared and never hear their song, the middle part is slurred. If they're exposed to wild "tutors" after they're a year old, they don't improve this middle part, so there must be a critical time for learning it during their first year. If hand-reared Wood Thrushes are exposed to tapes with only the middle part of the song, they end up singing a normal song, with the middle part matching the taped version.

Bird songs can carry us away with their beauty. But the more we learn about them, the more thrilling those songs become.

Male mockingbirds may have two distinct repertoires of songs: one used in spring and another used in fall.

BIRDS THAT MAKE SOUNDS
WITH FEATHERS

As a Mourning Dove flies past, you may hear a whistle — a sound produced by its feathers. The function is unknown, but some scientists speculate that sound may signal alarm when the birds take off. Male Mourning Doves also use their wings to produce noise when attracting the attention of females. They launch from a perch high into the air, flapping their wings loudly, then glide down as part of a courtship display.

Many other birds also produce sounds with their feathers. Ruffed Grouse make an explosive thunder as they burst into flight. Common Goldeneye wings make a whistling sound in flight that can carry more than a mile. American Woodcock wing feathers make a distinctive twittering sound as the birds fly.

In a territorial display, male Anna's Hummingbirds plummet through the air nearly vertically from heights of 66 to 131 feet (20–40 m). At the bottom of their dive, they make a loud squeaking noise, similar to a sound they make while singing from a perch. Using ultra high-speed cameras, researchers found that the birds actually produce the dive squeak when they spread their tail at the end of their fall, causing air to vibrate the feathers rapidly. The scientists were able to produce the same sound by blowing a stream of air over a hummer's outer tail feather.

Similarly, Wilson's Snipes use their stiff outer tail feathers to produce sounds when they display in spring. The snipe circles high overhead and then drops, speeding up to as much as 52 miles per hour in the dive, while spreading its tail. The vibrations of the two stiff tail feathers produce a peculiar hum as the wing beats send the air over the tail in pulses. The distinctive sound, *woo-woo-woo-woo-woo*, called *winnowing*, is audible from up to half a mile away.

Tropical manakins make an array of bizarrely wonderful sounds with their wings, from rapid firecracker-like snaps to tonal hums with their feathers. High-speed video shows that when a Club-winged Manakin flips its wings over its back, it knocks two specialized wing feathers together, creating a high-pitched tone as one feather slides against the other, like a bow on the string of a violin.

CHAPTER 7

Do You Know the Way to San Jose? Bird Migration

Since earliest times, people have wondered about the comings and goings of birds. Ancient Greeks explained their seasonal disappearance by a belief that they buried themselves under mud for the winter. People in the Middle Ages thought birds wintered on the moon. It wasn't until world travel became more prevalent that people began to comprehend patterns of bird migration, and even in the twenty-first century, scientists are still unearthing new mysteries about it. If they still have questions, is it any wonder that the rest of us do?

^v^v^v^v^v^v^v^v^v^v^v^v^v^v^v^v^v^v^v^v^v^v^v^v^v^v^v^v^v^v^v^v^v^v^v

The Mysteries of Migration

Q Why do birds migrate?

A Most birds that breed in the huge landmasses of the Northern Hemisphere feed on insects, fruits, fish, and weed seeds — foods that disappear or are covered with ice or snow in winter. These birds must migrate south to find suitable food in winter, but they return north apparently to capitalize on the rich food and fewer competitors during the breeding season.

Although many people assume that birds migrate to escape cold winter temperatures, birds can actually survive very cold temperatures, as long as they have enough food. Even tiny birds, such as Boreal Chickadees, can survive temperatures as low as –50°F (–46°C). They survive the long winter months by eating spruce seeds and insects they have stored in the crevices of trees.

∨ ∨

Q **How far do birds migrate?**

A It depends on the species. Some birds don't migrate at all, such as Florida Scrub-Jays, which remain near their area of birth for their entire lives. A few mountain species such as Sooty Grouse just move up and down the mountain with the seasons. Depending on where they live, individuals of some species, such as American Robins and Dark-eyed Juncos, may move less than a hundred miles.

Other species travel thousands of miles each year. Eastern Kingbirds that nest as far north as Canada's Yukon, Northwest Territories, Labrador, and Newfoundland migrate all the way to South America for the winter. Arctic Terns and many shorebirds

that nest on Hudson Bay or the Arctic tundra winter in south-ernmost South America, traveling 24,000 miles (38,625 km) each year between their nesting and wintering grounds. Arctic Terns may travel as far as 1,800 miles (2,897 km) between rest-ing and feeding stops during migration. Some Pacific Golden-Plovers migrate between western Alaska and the Hawaiian Islands, an overwater journey of more than 3,000 miles (4,827 km). Sooty Shearwaters have the longest migration on record: they travel about 40,000 miles (64,360 km) each year, traversing the Pacific Ocean in figure eights.

˅ ˅

Q **How do birds know when to migrate?**

A From the time they hatch, migratory birds apparently respond instinctively, growing restless as the day length and angle in the sun change in spring and fall. Ornithologists have termed this migratory restlessness *Zugunruhe.* Even indi-vidual migratory birds that have been hatched and reared in captivity experience this urge.

By responding to day length, birds arrive on their breeding grounds at the best average time, regardless of weather condi-tions wherever the bird has wintered. In fall, this restlessness helps ensure that the journey will take place while rich food resources are most likely available, or, in the case of birds that must cross large bodies of water without stopping, that they consume rich food supplies before starting out. Many people assume that their feeders will entice birds to remain too long, but *Zugunruhe* ensures they do not.

Although this migratory restlessness provides the urge to move, conditions from day to day may provide the impetus to send birds packing. In spring, Neotropical migrants usually wait for northerly storms to pass over rather than flying into headwinds. Birds wintering in the southern United States often begin moving north with good weather. Robins and geese seem to follow the 37-degree *isotherm* (the "line" visible on a weather map where temperatures average 37°F [3°C]), from which they can backtrack when a sudden deep freeze puts them in danger.

Spring arrival dates for these species can vary widely from year to year. Birds wintering in the tropics can't predict what weather will be like in the north, so their flights are more precisely tied to day length. One concern for them with continued warming trends is that insect emergences are now beginning earlier, coinciding with warm weather. Meanwhile, the Neotropical migrants arrive and begin nesting and raise their young at closer to the normal arrival dates, and now seem to have a poorer chance of capitalizing on the greatest insect abundance while raising their hungry young.

Goldfinches move south in winter following a pattern that seems to coincide with regions where the minimum January temperature is no colder than 0°F on average.

A BIRD'S GOTTA FLY WHEN A BIRD'S GOTTA FLY

Researchers study migratory urges by putting birds in an orientation cage, called an Emlen cage after the scientist who designed it. It is a funnel shape in which birds can stand upright. Birds that are in a migratory state try to escape from the cage by moving from the bottom area up the sloping walls. Early models had an ink pad on the bottom, and birds would leave footprints on the funnel walls as they tried to fly up. Newer versions may be lined with typewriter correction paper or another material so that the birds leave scratches as they move about.

These marks are distributed fairly randomly by non-migratory birds or birds not in a migratory state, but show strong directionality in migratory birds undergoing *Zugunruhe*. Some modern cages also have perches fitted with microswitches that register a signal whenever a bird alights. These cages indicate how birds undergoing *Zugunruhe* are much more active than others, and also show preferred directions and length of time spent traveling.

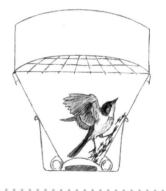

Q How do migrating birds know where to go and how to get there?

A Some species, such as cranes and geese, learn their migratory routes from their parents. They follow their parents on their first flight south, and sometimes for part of the return trip north in spring, and then are on their own. Most birds, however, cannot depend on their parents to lead the way. Young hummingbirds and loons migrate days or weeks after their parents have left, and yet the young birds know which way to head, how far to go, and when to stop. The more we study the intricacies of orientation and navigation, the more miraculous it seems.

In some cases, birds head in one direction for a certain distance and then change course. For example, Blackpoll Warblers start their migration in an easterly or southeasterly direction from Alaska and northern Canada toward the Atlantic coast. They bulk up, often doubling their weight, and then strike out over water on a journey that can take from 36 to 88 hours without a single break for food or rest, traveling 1,500 to 2,200 miles (2,414 to 3,540 km) nonstop until they reach South America. They start out over the ocean heading southeast, aided by northwesterly winds. As they approach the Tropic of Cancer, they start encountering the northeast trade winds, which deflect their flight to the south or southwest and provide favorable tailwinds as they make the final push toward South America.

In this case, favorable tailwinds may help send the birds in the right direction, but they can use other navigational aids as well. Research scientists are teasing out the environmental cues long-distance migrants use, including the earth's magnetic

ADOPTIVE PARENTING

Beginning in 2001, a population of captive-reared Whooping Cranes have been learning their migratory route by following an Ultralight aircraft driven by crane-costumed handlers that the birds are imprinted on. The birds follow the tiny plane from Wisconsin to Florida their first autumn, then make the return trip and subsequent journeys entirely on their own. Eight years after the first release, some of the birds are starting to breed, and ornithologists hope that soon this introduction will be providing a self-sustaining population of this endangered species.

fields, the movement of stars in the sky, and polarized light patterns. Recent research has discovered a neural connection between the eye and a part of the forebrain that is active during migrational orientation. This suggests that the visual system is involved in the birds' ability to sense the earth's magnetic field. Some species may also use visual landmarks, such as mountains or rivers below them.

Q How do birds prepare for long migrations?

A As days shorten at summer's end, photoreceptors in their brains trigger hormonal changes that stimulate many birds to molt into new feathers that will stand up to the rigors of a long flight. Their hormones also trigger a huge appetite, and they start eating voraciously, gaining significant amounts of weight. Many insectivorous species supplement their diet with fruits, grains, and other items that can be converted to body fat, which birds burn efficiently for energy. These hormonal shifts make birds increasingly restless, especially at nighttime. Suddenly, one day it's time to go!

Many of the best places for witnessing migration are along coastlines and bluffs. Watch for hawks and other birds that ride on thermal air currents. Warblers, thrushes, and other songbirds that cover long distances by night are more likely to be down in the vegetation feeding and resting. You can often find them by listening for chickadees, because when warblers and vireos are passing through an unfamiliar area anywhere in North America, chickadees allow them to join their feeding flocks. Chickadees know where the best food is and where predators are mostly likely to be lurking.

Q I've noticed that the male Indigo Buntings in our area look so bright and shiny in the spring, but by the end of the summer, they are drab and dull. What happens to them?

A Late every summer, after breeding, male Indigo Buntings molt out of their bright blue feathers into drab brown ones. They keep these feathers through their migration to the tropics and during most of the winter, and then in late February or early March they molt into their bright blue feathers once again just before returning north.

Bird feathers are wonderful for protecting their bodies from extreme temperatures, rain, wind, and too much ultraviolet light, but over time feathers grow frayed and parts break off. Molting provides a great way of replacing them before their general wear and tear cause problems. In the case of Indigo Buntings, the bright blue plumage of males is very useful when they're trying to attract mates and defend their territories but is not so helpful in winter, so the feathers they grow at that time make them less conspicuous.

By the way, the blue in their feathers is due to the way the outer layers of cells in the feathers reflect light, not due to any pigments. If you were to grind up a few male Indigo Bunting feathers, you wouldn't see a trace of blue — the feathers are pigmented to be dull grayish brown. The blue color is most intense when sunlight is bouncing off it, and least intense when the bird is backlit or in low light.

HITCHING A RIDE

I've often been asked if hummingbirds hitch rides on the backs of geese while migrating — a rather charming concept but definitely a myth. So where did this idea come from? Once I saw a Ruby-throated Hummingbird that was perched on a wire become suddenly agitated. I looked up in the direction it was looking to see a Bald Eagle flying high above. The eagle had apparently crossed into the hummingbird's column of defended airspace, and the hummer took off straight for it.

When it reached the eagle, the hummer started dive-bombing it on the upper back and nape, up, down, up, down, like a little avian yo-yo. Eventually, the eagle must have crossed out of the hummer's defended airspace. At that point, the little guy flew back down to his wire, chirping animatedly like he'd chased the big bruiser away.

Hummingbirds will attack virtually any large creature that enters its feeding territory. I wonder if someone once saw a hummingbird dropping down or flying toward a goose and thought it was hitching a ride, rather than maneuvering for an attack. I did read of one account by a hunter who shot a goose that had a dead hummingbird tangled in weeds on its back. If true, that's hard to explain, but it may be because the hummer got tangled while doing a dive-bomb. I think people just find it hard to believe that tiny hummingbirds can migrate so far — it may just seem more believable that they hitch a ride!

In fact, though, hummers can even fly across the 600-mile (965 km) Gulf of Mexico on their own. We know this because specially licensed bird banders capture, weigh, and band hummingbirds in southern Texas, Louisiana, and Mississippi before they cross the Gulf. Other researchers capture the hummingbirds when they arrive on the Yucatán Peninsula and weigh them again. The average difference in weights fully accounts for the birds flying nonstop over the Gulf of Mexico on their own power.

Q **Why do I see some species every winter, but other species only every few years?**

A Some food and nesting resources are predictable and found readily year after year, but some vary greatly. Robins can return to the same yards every spring for nesting, and except in extreme droughts can count on a fairly regular supply of worms for feeding on and mud for nest building. But in winter, the trees of one area can be rich in fruits one year and barren the next. So they wander widely in winter. Many finches depend on conifer seeds during the winter. One year conifer cones may be abundant in one area, and the next year birds may have to travel hundreds of miles to find the same food. The birds that move about feeding on these unpredictable food sources are called "irruptive."

❧ ❧

Q **Why do some hawks migrate and some stay through the winter?**

A Different hawks specialize on different prey. Those that eat rodents in large fields can either head for South America and find rodents in large, summery fields down there, as Swainson's Hawks do, or stick it out in the north, as Red-tailed Hawks do, grabbing animals in large fields whether they're peeking out from the snow or running across a road.

Peregrine Falcons that specialize on shorebirds often fly all the way to South America; those that learn to catch city pigeons can spend the entire winter in the north. Some individual

Cooper's Hawks have learned to exploit bird feeders for easy prey. They remain north while other individual Cooper's Hawks head to Central America to feed on warblers, tanagers, and orioles.

❧ ❧

Q **Every summer most of my hummingbirds disappear for a few weeks and then suddenly reappear in large numbers. What's going on?**

A Adult male hummingbirds aggressively defend their territories, and if your yard is within the territory of one, he may drive all other male hummingbirds away during the nesting season. If you have a nesting female nearby, she will visit your feeder only periodically, spending most of her time incubating her eggs.

After the eggs hatch, she usually concentrates her feeding at flowers that supply tiny insects as well as nectar. Insects contain the protein that her nestlings need in order to grow, and she's regurgitating a slurry of the nectar and insects she's eaten to feed them. Once the young have fledged, she continues feeding them for several days until the fledglings have mastered getting their own food.

At this time, she may bring them to your feeders to teach them how to take advantage of this easy food supply, too. This is also when males stop defending a territory and begin migrating, with adult females soon following. Many of the hummingbirds that suddenly appear are actually migrants from farther north, just passing through.

Taking the Night Flight

Q **I've heard that many songbirds migrate at night. Why?**

A There are several major advantages to nocturnal travel.

▶ Temperatures are cooler and the air is moister, protecting them from overheating or dehydrating.

▶ Winds are often lighter, helping them to conserve energy.

▶ Songbird-hunting hawks are active by day and nocturnal owls don't chase prey in midair, so migrants are safer from predation at night.

▶ Since songbirds require daylight to see and capture their food, nighttime travel allows them to spend their days fueling up for the next leg of their journey.

ᵛ ᵛ

Q **How do birds see where they're going when they migrate at night?**

A They navigate by sensing magnetic fields and by using the stars to orient themselves in the right direction. Birds, like us, have "rod" cells in their retinas, providing some limited nocturnal vision. And, also like us, they have an easy time seeing the stars. But they can't see other objects in the sky or below them very well, particularly on moonless and cloudy nights.

Researchers have long suspected that some migratory birds can use the earth's magnetic field to orient. One classic study in the 1960s involved homing pigeons wearing tiny metal helmets. Half the helmets were magnetic; the other half were the same weight and size but not magnetic. On sunny days, all the birds found their way home easily. On cloudy days, the birds sporting magnetic helmets didn't orient or find their way back, at least not until the sun came out again.

More recently, researchers have discovered deposits of magnetite in the nasal tissues of several migratory species, including Bobolinks and White-throated Sparrows, and have learned that some neurons in visual centers of the brain respond to changes in the magnetic field. In darkness or red light, these birds may become disoriented, but in white, green, or blue light their sense of direction is normal.

In 2004, a European team studying the Garden Warbler discovered light-sensitive pigments in the retina that influence how the bird's sensory cells react to the magnetic field. It's possible that the curvature of the retina and the position of the magnetite in the nares (nostrils) may create some sort of pattern as they respond to the magnetic fields, perhaps producing a visual color shift when facing north or south but not east or west. This would explain why birds grow disoriented in some colors but not others, but we don't know if the birds' perception of magnetism is through vision or some other sense that we mere humans can't even imagine.

Q How do scientists know that birds use the stars for navigation?

A Fascinating research projects conducted in planetariums were begun in the 1950s, when European scientists Franz and Eleanor Sauer discovered that Garden Warblers fluttered away from the projected North Star, even when the projection was rotated to place the North Star in another direction. During the 1960s, American scientist Stephen Emlen showed that birds don't recognize one particular star but actually the pattern of stars revolving around the one fixed star. He raised young Indigo Buntings in a planetarium, under a projected sky with stars revolving around Betelgeuse. During the birds' first autumn, they oriented to fly away from Betelgeuse.

Apparently young birds spend part of the nighttime awake, gazing at the night sky. They seem to notice that the stars move in a circular pattern, and instinctively know that the one fixed star that never moves is "north," and they can find it by learning the pattern of the stars around it so they can figure out where north is even when the sky is covered with patchy clouds.

Millions of migrating birds strike high-rise buildings every year. What can we do? Toronto's Fatal Light Awareness Program (*FLAP.org*) patrols downtown Toronto in early morning to rescue live birds and collect dead ones. They urge people to turn off lights or close drapes during migration. Chicago, Minneapolis, and San Francisco have similar programs.

Following Directions

Q Do birds migrate directly north and south? How do they compensate for crosswinds?

A Few birds fly directly north or south. Many species, such as Blackpoll Warblers, have adaptations allowing them to capitalize on specific feeding or flying opportunities along their route. Many of these birds start out heading east-southeast toward the Atlantic Coast, and then take off over the ocean toward South America, aided by trade winds. To aid their long flight, their primary wing feathers are longer than those of warblers, such as Pine Warblers, that make shorter flights.

Ruby-throated Hummingbirds fly generally south until they reach the Gulf Coast, and then take advantage of abundant food in the Mississippi Delta to build up their fat reserves. Then many of them take off over the Gulf, flying nonstop to the Yucatán. But there is evidence in both cases that some individuals hug the coastline rather than making such a long overwater flight.

The flight direction of migratory birds in general depends on where their final destination is, whether the stars are visible, and a lot of individual factors. For example, thrushes have a very strong sense of direction. From a given departure spot, though, each bird takes a different heading. They're all moving in general toward their species' winter range, but a winter range can be large — it wouldn't be very good for all the birds to end up in exactly the same spot!

Q Since they can fly away, why do some birds stay in extremely cold places for the winter?

A The average annual adult survival rate of year-round tropical residents is 80 to 90 percent. The average annual adult survival rate of migrants is about 50 percent. And the average annual adult survival rate of temperate zone year-round residents is only 20 to 50 percent, because of the difficulties of surviving severe winters. Why don't these northern residents head south to improve their odds?

The evolutionary game is won not by the birds with the longest lives but by those who produce enough young to replace themselves. Year-round residents can select the best territories long before migrants arrive and can work out territorial differences well in advance of the breeding season so they aren't depleted by the rigors of both a long journey and territorial battles right when they start nesting. They can also get an earlier jump on nesting, allowing them more time to re-nest if their first attempt fails.

˅ ˅

Q Migrating seems so dangerous and the birds must use so much energy to fly such great distances. Why do they do it?

A Migration in birds probably evolved many times, for many reasons. The northern temperate and boreal zones are vast and rich with insects; they have fewer marauding insects such as army ants, and fewer poisonous snakes and

BLOWN OFF COURSE

🐦 **Although the mechanisms of bird orientation** and navigation are amazingly intricate, some storm systems are powerful enough to send migrating birds far off course. Individual North American songbirds, including several species of warblers, Dark-eyed Juncos, White-throated Sparrows, Scarlet Tanagers, and Rose-breasted Grosbeaks, have been carried all the way across the Atlantic, especially to England and Ireland, frequently enough to be included in some field guides to the birds of Europe. A few Siberian birds appear with some regularity in North America, especially Wheatears and Bramblings. And some birds seem to have a compass that can become so disrupted it entirely reverses their migration, such as Fork-tailed Flycatchers from South America that have turned up dozens of times in North America. To see the migration routes of warblers and other birds, go to *http://bna.birds.cornell.edu/bna/species/431/articles/migration*.

spiders than are found in the tropics. The northern latitudes also have significantly longer summer days than tropical zones, providing more hours for feeding young each day and significantly reducing the amount of time vulnerable young remain in the nest.

Some birds breeding in even northernmost North America may well have originated as tropical residents who, to escape competition and take advantage of huge insect population explosions in the north, started heading to the vast northern landscape to breed, retreating "home" to the tropics again after the breeding season. A full 78 percent of all northern migrant species, including flycatchers, swallows, vireos, wrens, and orioles, have close relatives in the same genus or even the same species that are year-round Neotropic residents.

Some non-migratory populations have probably become migratory as conditions changed for them. One example that we've been able to observe firsthand: non-migratory House Finches from the Southwest were released in New York in the early 1940s. Within 20 years, some of these birds had begun migrating to the Gulf States for the winter and returning north to breed. Other individuals remain year-round in the Northeast. In just two decades this species has become a *partial migrant.*

∨ ∨

Q I've heard you can tell if birds are migrating on a particular day by looking at a weather map! How does that work?

A It's true. Weather radar images show where radar beams have been "reflected" as they sweep the atmosphere. They're useful for showing weather conditions because the beams are reflected by precipitation and the water vapor in clouds, but they can also be reflected by swarming masses of birds or insects.

In the early days of World War II, British radar operators noticed mysterious, ethereal shadows drifting across their screens. They weren't associated with weather systems and so the radar technicians nicknamed them "angels." In 1958, a New Orleans high school student named Sidney Gauthreaux, realizing that these "angels" were really the radar reflections of swarms of birds, started scrutinizing radar images. As a Louisiana State graduate student, he worked with radar images to document the existence of massive trans-Gulf migrations.

In the late 1980s, Gauthreaux started examining archival radar images and made a disturbing discovery: major bird movements over the Gulf had declined by nearly half since the 1960s.

Next Generation Radar (NEXRAD) made studying bird migration much easier. The Air Force started using it to avoid collisions in their Bird Aircraft Strike Hazard program. Graduate students took stunning images of giant expanding aerial doughnuts, which they found to be thousands of Purple Martins radiating from critical roosting sites each morning.

Now it's easy for anyone with access to a NEXRAD weather map on their computer to see birds take off on migratory movements at night or alight in the morning, if you know how to interpret the mystifying patterns. You can learn how at *www. virtual.clemson.edu/groups/birdrad/*.

THE DANGERS OF MIGRATION

Fall migration, when many birds fly over the Gulf of Mexico, takes place during hurricane season. Birds can be killed outright by high winds, hail, blowing debris, falling trees, and so forth. They can be blown too far off course to survive. Birds passing through a devastated area after a storm may have trouble finding feeding resources, or may succumb to pollutants released in hurricane-related oil, gasoline, and other toxic spills caused by flooded automobiles, households, oil refineries, chemical manufacturers, and other sources.

Most migrants don't encounter hurricanes and avoid damaged areas. But there are plenty of other dangers out there. Communications towers in their flying space may kill as many as 50 million migrating birds per year. Occasionally in foggy or stormy weather when birds are flying over open water in the Gulf or the oceans, they are attracted to the lights of a lighted vessel and dash against the windows, injuring or killing themselves. And if they survive the dangers of the night, come morning, they have to negotiate whatever habitat they find themselves in.

Cities can be treacherous. An estimated 500 million to one billion birds are killed each year in the United States in collisions with windows. Many of these deaths take place at lighted high-rise windows at nighttime, and many take place in the morning against plate glass windows and doors at ground level, when birds find themselves in a heavily urbanized area where the only vegetation is inside hotel lobbies and solariums. Habitat degradation, especially along coasts and shorelines, can make obtaining food difficult for migrants right when they need it most. As a wildlife rehabilitator, I cared for several loons, grebes, and rails (shy little marsh birds) that had crashed onto sparkling wet pavement where a wetland existed the previous year.

One researcher found that apparent mortality rates of Black-throated Blue Warblers were at least 15 times higher during migration than during the breeding and winter resident periods; more than 85 percent of apparent annual mortality of this species occurred during migration!

You can see a map of where any wintering species were reported in mid-February in the United States and Canada during any winter since 1999 on the Great Backyard Bird Count map; the website is *www.birdcount.org*.

What Birds Do in the Winter

Q **Do birds nest again once they reach their wintering grounds?**

A No. Northern birds spend winter eating food in the tropics and, sometimes, molting into new feathers. Raising young takes tremendous amounts of energy, and few migratory species can afford to do that for longer than a fairly brief period of time in their annual cycle.

Canada Geese were introduced into Great Britain as early as the 1600s. Four centuries later, the geese on the British Isles are still non-migratory. Canada Geese were introduced into Sweden during the first half of the 1900s. These geese have become migratory, many headed to Scania, in the southernmost tip of the Scandinavian Peninsula, and to the East German Baltic coast.

Q **What do birds do while they are "on vacation" in the tropics? Do they just loaf around and eat?**

A Many birds, males and females alike, use as much energy defending a winter territory as they do a breeding territory. There are great advantages to knowing every inch of a familiar plot of land for finding food, hiding from predators, and having safe roosting places. There are many dangers in the tropics, and birds must compete with a tremendous number of other birds for resources.

Some birds, such as Eastern Kingbirds, that specialize on a high-protein insect diet in the north while producing eggs and feeding young may switch to a fruit diet in winter. Fruit-eaters such as these may be highly territorial in summer, but on their wintering grounds peaceably associate in flocks, wandering widely in search of new food supplies.

Whether they defend a winter territory or associate in a flock, wintering birds must recover from their arduous autumn migration and get back in condition to migrate north all over again when the time comes.

∨ ∨

Q **Do any Central or South American birds spend their winter up here in North America?**

A A few individual southern birds wander north after their own breeding season, but those are anomalous cases. About 220 to 240 *austral migrants* (birds that breed in the Southern Hemisphere and migrate north for their winter) breed in temperate South America and winter toward the Amazon basin. Most austral migrants do not winter as far as the tropics — only 32 species reach Amazonia and 14 more winter north of the Amazon basin. This, by the way, compares with about 420 species that breed in temperate North America and winter in the Neotropics.

Robins are supposed to be harbingers of spring but I see them all winter in Minnesota! How is that possible?

A Robins switch their diets from primarily worms and insects in spring and summer to primarily fruits in fall and winter. They can survive as far north as they can find a consistent supply of food. Robins are interesting migrants — rather than a north–south route, they wander more loosely in search of their winter diet of fruits and berries, which are unpredictably abundant in some places and sparse in others.

Those robins that remain in the far north tend to be males who get the advantage of arriving first and claiming the best territories in spring, if they survive the winter. They will move on if they deplete the food supplies, and some succumb to bad weather and starvation. So the genes for longer distance migration remain alive in the population, too. Major spring migratory movements of robins follow the thaw pattern when earthworms first appear, about where temperatures are starting to average 37°F (3°C). But in many northern areas, wintering robins are around long before the "first robin of spring" arrives.

How do you tell that first robin of spring from the last robin of winter? If they're feeding in flocks in fruit trees, they're still exhibiting winter behaviors. When they run on lawns feeding on worms and start singing, they're showing spring behaviors; usually spring robins are seen as individuals, pairs, or fighting rivals. In the changeable conditions of early spring, a robin may exhibit spring behaviors one day and revert to winter behaviors when a late blizzard blows in.

CHAPTER 8

The Birds and the Bees: How Birds Reproduce

Since earliest history, eggs have served as evocative symbols of new life and rebirth. We humans turn to the avian world to describe our own "broodiness," our urge to save up a "nest egg," and how, when our children leave, we have "empty-nest syndrome." When June, and eggs, are bustin' out all over, we're filled with joy and wonder. And questions.

∧∨

The Mating Ritual

Q The song says, "Birds do it. Bees do it. Even educated fleas do it." But what exactly is it that birds "do"?

A Before mating, birds often court one another with ritualized displays. For example, a pair of Whooping Cranes will face each other and dance, leaping into the air, bowing, and flapping their wings. Red-tailed Hawks may grasp one another's talons in flight and spiral together toward the ground. Male songbirds sometimes present their mate with a choice insect or berry. Mallards bob their heads up and down, the pair moving their heads in opposite directions so that one is bobbing up while the other bobs down.

When ready to mate, in most cases, the male bird alights on the female's back. The birds twist their tails so that their cloacae (the common chamber where the urinary, intestinal, and genital tracts open) meet. The sperm pass from the male's cloaca to the female's. As the fertilized egg works its way down

234

the female's oviduct, the cells along the way secrete the proteins that make up the albumen, and then secrete the calcium that will form the shell. Usually by early the next morning, the female lays the egg. Females have one functional ovary and in most species ovulate once every day or two, so the laying period lasts at least as many days as there are eggs in a clutch.

During the days or weeks between starting nest building and finishing laying a clutch of eggs, a pair of birds may mate several or many times, and sperm can remain viable inside a female often for more than a week. If her mate is killed, she may be able to raise young on her own, though normally she replaces her lost mate with another male within a day or so. In recent decades, ornithologists have been discovering that many species have "extra pair copulations." Even when female birds are socially bonded to their mate, they sometimes mate with another male once or twice, and often the young birds in a single brood have more than one father.

Q Why do birds have such elaborate and varied courtship rituals?

A Courtship rituals are a form of communication, enabling birds to signal their willingness to mate. They also give the birds an opportunity to assess their partner. A female bird invests a great deal, physiologically and energetically, in producing eggs, incubating them, and raising the young. Courtship displays can help her select a mate who is most likely to produce healthy young. She may look for clues about his health, vigor, or ability to provision the young, based on his appearance, his display, or his song.

For example, a male bird may show off his brightly colored plumage because bright colors indicate his health or ability to find good food. A male Snail Kite offers his mate a stick or a snail, perhaps a sign of his ability to provide materials for a nest and his superior hunting skills. Male songbirds may sing repeatedly to advertise their vigor or experience. Female Northern Mockingbirds may prefer males that sing the most song variations. Since older males typically sing more songs, a larger repertoire may indicate longevity and experience in raising young.

The male Eastern Bluebird displays at his nest cavity to attract a female. He brings nest material to the hole, goes in and out, and waves his wings while perched above it. Only the female builds the actual nest and incubates the eggs, but he shares the work in feeding the nestlings.

Q Why do Blue-footed Boobies have blue feet?

A Blue-footed Boobies, fish-eating relatives of pelicans, breed in tropical and subtropical islands in the Pacific Ocean, most notably the Galápagos. They have large webbed feet ranging in color from pale turquoise to a deep aquamarine. Males flaunt these feet in courtship dances, holding them up and stamping them on the ground. Males with the most brilliant feet are the most attractive to females. Booby foot color is brightest when birds are well fed on nutritious fish — in captivity when deprived of a good diet, the foot color grows noticeably duller within 48 hours. So this is an excellent cue for females to use to determine a male's ability to find food, a direct signal about how well he will provide for young. Interestingly, if a scientist dulls the color of a male booby's feet with paint after his mate has laid her first egg, her second egg will be small, perhaps to reduce the food resources necessary to raise that chick.

Although the male Blue-footed Booby dances, showing off his blue feet conspicuously in a courtship ritual, the blue foot color is present on both males and females throughout the year, not just during the breeding season. Both sexes use their feet to incubate their eggs and to keep the chicks warm, but the color doesn't enhance those functions.

❮ ❮

AMAZING COURTSHIP DISPLAYS

Western Grebes, grassland grouse, Calliope Hummingbirds, manakins, tropical wrens — the vast array of spectacular courtship displays fills humans with wonder even as they set bird libidos on fire.

Many people have seen film footage of a pair of Western Grebes (elegant, long-necked, black-and-white diving birds from the American West) dancing in synchrony on the surface of a lake. In these spectacular dances, called the "rushing display," two birds turn to one side and lunge forward with their bodies completely out of the water and run rapidly across the surface side by side. Occasionally one or more other birds join them. After "rushing" back and forth for 5 to 20 minutes, the birds lower their wings and dive into the water. They don't call during this dance, but the pattering of their feet on the water is distinctive and thrilling to witness.

Greater Sage-Grouse, Greater and Lesser Prairie-Chickens, and Sharp-tailed Grouse — all large chicken-like birds of the American West through the northern plains — have wonderful displays. Males gather together on a breeding ground called a *lek,* inflating brightly colored air sacs, stomping their feet, erecting some body or head feathers, and opening their wings. Females wander through and choose the males with the best displays to mate with. These grouse form no pair bond: the females build their nests and care for eggs and chicks entirely on their own. They'll fly in from several miles away to a lekking ground to choose the finest males.

Many species of manakins, from Central and South America, have amazing displays, from the "moonwalking" Red-capped Manakin (see page 162) to the White-collared and Orange-collared Manakins. In these, males clear a patch of ground as a lek and leap back and forth between thin upright bare sticks, giving a loud wing snap with each jump. When a female is present males jump together, crossing each other above the bare display court. They erect their throat feathers to form a beard during displays.

Many male hummingbirds have spectacular courtship flights to attract females. For example, the male Calliope Hummingbird, the tiniest North American species, living in the mountainous West, makes a spectacular dive: climbing to about 100 feet (30 m) and then plunging at high speed and braking abruptly, then climbing again, following a U-shaped trajectory. During this, the flight feathers make a buzzy *hum*.

Many tropical wrens have an auditory rather than a visual courtship display, singing elaborate duets so seamlessly integrated that it can be difficult to discern that more than one bird is singing.

Q I was watching birds at the feeder and noticed a male Rose-breasted Grosbeak feeding either its mate or a fledgling, which was perched very close by waiting for the male to bring it seed. Why wasn't the female or fledgling going to the feeder itself?

A During courtship, males sometimes offer food to females — the females may use this to determine how good a provider each of her suitors could be. After the babies fledge, it takes them a while to figure out what foods are appropriate to eat, and they may still be clumsy fliers. So it is probably safer for a young bird to sit in the relative shelter of a shrub while the adult flies through the open to and from the feeder bringing it food.

^v^v^v^v^v^v^v^v^v^v^v^v^v^v^v^v^v^v^v^v^v^v^v^v^v^v^v^v^v^v^v^v^v

Birds of a Feather

Q Why do robins lay blue eggs? Is it possible to tell males and females apart?

A No one knows why robins lay blue eggs, but the eggs of most thrush species are blue. The blue doesn't seem to camouflage the eggs, but it doesn't make them particularly conspicuous either. Pigments usually add some structural strength to tissues, and this particular blue pigment may give the eggshells just the right thickness and hardness to protect the embryo while allowing it to finally break out.

Robins flock year-round except during the breeding season. The mother remains at or near the nest site all the time when she has eggs or nestlings, but after the first brood of young fledge, the father takes them at nighttime to a communal roost. When he leaves them after 12 or 14 days to help raise the next brood, they remain together and with other young robins.

After the pairs have finished raising young, they also join these flocks. During fall and winter, robins can sometimes be found in flocks numbering in the thousands, ten thousands, or even hundred thousands, though most flocks seem to number a dozen or two. These flocks break up in spring when males and females appear restless and territorial squabbles erupt more frequently.

The differences between male and female robins are very subtle. Females are overall paler than males, especially on the head, which is almost black on males. The white eye crescents and throat striping are a bit more dramatic on males, too. But when robins from other areas join together, especially in winter flocks, sometimes you'll notice that bright females from one area are as intense as dull males from another area. It's easiest to see the differences between a pair that you're regularly watching, especially during the breeding season. Also, if you hear a robin singing (not just making the *peek!* or *tut-tut-tut* calls), it's a male for sure.

˅ ˅

BOYS AND GIRLS TOGETHER

🐦 **The male and female of some species of birds,** especially those that mate for life, such as jays, crows, ravens, cranes, eagles, geese, and swans, are identical or differ only in size. In other species, the male and female may bear dramatically different plumage, such as cardinals, waxwings, finches, and many ducks. This is called sexual dimorphism.

Females are often fairly cryptically colored, which helps camouflage them on the eggs, though male Rose-breasted Grosbeaks, despite their brilliant color and pattern, also incubate. Males are bright, which helps them attract a mate and aids in their territorial defense. For example, female Red-winged Blackbirds and female Baltimore Orioles tend to prefer males with the most brilliant colors, and these males tend to defend the best territories.

For many birds, the name of the species really describes only the male. Female Scarlet Tanagers are never scarlet, female Red-winged Blackbirds do not have red wings, and female Ruby-throated Hummingbirds do not have ruby throats.

Q Do male American Goldfinches change color in winter in the Midwest? Or do they migrate and leave the females here?

A Winter male American Goldfinches, like females, are dully colored. The difference between winter and summer plumages is the most striking in its family, and the spring molt is unique among finches. Across the continent, migratory activity peaks in mid-April to early June during the spring molt, and in late October to mid-December after the fall molt. There is a great deal of variation in individual birds — some may migrate some years and not others. Goldfinches associate in flocks that include both sexes, and individuals of both sexes migrate or remain residents in the Midwest as they please.

^v^v^v^v^v^v^v^v^v^v^v^v^v^v^v^v^v^v^v^v^v^v^v^v^v^v^v^v^v^v^v^v^v

Location, Location, Location: Defending Territory

Q **In my local park, I saw two mockingbirds fighting, but I couldn't tell what they were fighting over. It seemed like there was plenty of space for both of them.**

A When it's time to nest, many birds establish and defend territories, and fight for mates. To successfully raise young, birds need a safe place for their nest and adequate food to feed their family. Many bird species defend an area around their nest site, chasing away other birds that might compete with them for food. They may also chase intruders away to prevent rivals from mating with their partners. A Northern Mockingbird's territory may range in size from about 1 to 6 acres (0.4–2.4 ha). Some songbirds have relatively tiny territories — Ovenbirds and Song Sparrows may require less than half an acre. But songbirds that require specialized food or food from

243

higher up the food chain require more space. Scarlet Tanagers can require 5 to 30 acres (2.0–12 ha) or more, depending on the quality of the habitat. Loggerhead Shrikes may require 75 acres (30 ha) or more, and Gray Jays average about 250 acres (101 ha).

In nonsongbirds, too, the size of a territory is affected by the abundance and distribution of food and nest sites. Bald Eagles in sparsely vegetated Saskatchewan defend territories at least 1½ square miles (3.9 square km) in size; the minimum size of their territories on Kruzof Island in Alaska, where food resources are more abundant, is only ⅛ square mile (0.5 square km). Some eagles may fly quite a distance from their nest to fish, so if there are abundant fishing waters in one area, eagles can be somewhat concentrated where large trees provide good nest-ing. Once an eagle catches a fish, other eagles some-times aggressively fight to take it, so even though they can share fishing areas, they need at least some space between one another. Red-winged Blackbirds, by contrast, vigorously defend their nesting territories but feed peaceably, almost shoulder-to-shoulder, away from the nesting area.

v v v v v v v v v v v v v v v v v v v v v v v v v

Q **How many pairs of wrens can live peaceably in my backyard? Are they extremely territorial? I'd like to put up several wren houses.**

A Wrens are *extremely* territorial. You'll get one male at most, but he may build stick nests in each of your wren houses, and may attract more than one mate. Each female will choose one box to nest in.

˅ ˅

Q **Do all birds defend territories?**

A No. Some birds are extremely sociable even during the nesting season. For example, thousands of Cliff Swallows may build their gourd-shaped mud nests side by side in a single dense colony. One reason they live in close proximity is because the specific sites where they can find mud of the right consistency and suitable structures on which to build their nests are limited. Each pair defends the area right where they're building, but once the nest is built, they don't seem to mind if other swallows perch on the nest itself as long as they don't enter. After they finish building, they defend just the area immediately beneath their nest to prevent other birds from building a nest that might block the entrance to their own nest.

Great Blue Herons also nest in colonies, constructing their stick nests in a stand of dead trees. Adults may fly several miles to and from good fishing spots, and they tend to be rather territorial around their fishing grounds, but at the nest sites they are gregarious.

Emperor Penguins don't defend territories either. They stand side by side while incubating their eggs and raising their chicks. What do Cliff Swallows and Emperor Penguins have in common? They can both find areas with concentrated food distant from their nesting site. Cliff Swallows chase after swarms of insects. Penguins leave the nesting area to fish in the ocean. Both of these sources of food would be hard to defend.

By nesting closely together, the birds can benefit from better protection from predators; while some birds are off searching for food, other adults are nearby to detect and drive away marauders. Great Blue Herons build the kinds of structures that eagles or osprey like to take over, but it's difficult for these large raptors to take over a nest in a heronry, and the presence of so many large herons tends to keep the raptors away. Penguins get a special benefit from nesting in tight colonies during cold weather: They can crowd together, keeping each other warm.

Whooping Cranes in northern Canada defend territories that vary widely in size, from just a half of a square mile to 18 square miles (46 square km), depending on the resources available to them.

Q Our dog, a 10-pound rat terrier/miniature pinscher who doesn't bark, is getting swooped quite frequently by a couple of aggressive blackbirds. They have actually nicked him while he's in our yard. We have a number of evergreen trees on our property. Is it possible there is a nest? We've never had the issue before. We have also noticed, as have our neighbors, that we have at least one hawk in the area. If it's a nesting/protecting thing, how long is it going to last? Our dog is terrified!

A I suspect the blackbirds are Common Grackles, which nest in conifers. They may well have been there in past years, too, but this year the hawk may well have added to their stress levels. When your dog was the only perceived danger, they could go about their business while merely keeping an eye on it. Now that there is a much huger danger to them and their young, their rising stress levels may be making them act more aggressive in general. If it's a Cooper's Hawk, they don't dare mob it, so they're taking out all their frustrations on your poor little dog.

Grackles incubate their eggs for 11 to 15 days, and then the nestlings remain in the nest for 10 to 17 days. So, worst-case scenario, they'll be harassing your dog for a full month. But they don't re-nest, so when the babies at long last fledge and the families move on, your dog will be left in peace.

WHAT'S IN A TERRITORY?

A bird's diet and nesting habits determine what kind of territory it will defend. A suburban robin prefers moist lawns with earthworms and berry-bearing plants to provide food, and good tree limbs or house eaves for a secure nest site. A Barn Swallow looks for a bridge, culvert, barn, or other building that provides struts, eaves, or other supports for its heavy mud nest, near a pond or open field rich with flying insects.

Both species require mud for nest construction, so the best territories for each will also contain wet, muddy shorelines or good puddles. In good swallow areas, insects are far more abundant than the birds require, so many may nest close to one another without it hurting each other's ability to find food for their own young. Robins could much more easily deplete the supply of worms and other food on their territories if other robins nested too closely, so they defend their territories far more vigorously than do swallows.

Most hummingbirds seem to prefer territories with plenty of spider webs and lichens because they use these materials to build their nests. An ideal territory might also include nesting sapsuckers. Before flowers open and produce nectar in spring, Ruby-throated Hummingbirds can drink sap from wells in trees drilled by Yellow-bellied Sapsuckers. Birds that nest in cavities but do not excavate their own, such as bluebirds and wrens, search out these structures in suitable habitat that will supply their food needs.

Males of a few species, including tropical manakins, birds-of-paradise, bowerbirds, and cocks-of-the-rock, defend territories just so they can display to females there. Dozens or even hundreds of male Sharp-tailed Grouse or other grassland grouse species, including prairie-chickens, gather on large display areas called leks. These spots may look like any others to the human eye, but the grouse may return to them year after year, ignoring other similar-looking spots nearby.

At the lek, males display by inflating colorful air sacs, erecting their tails, and stomping their feet. Competing males, especially those on the periphery of the lek, may come to blows, pecking, biting, feather-pulling, wing-beating, and clawing at one another, especially when females enter the lek. Females may visit a lek several times during the early nesting stages, but they build their nest and raise their chicks completely independently of the males, sometimes more than a mile from the lek.

SEE ALSO: *page 238 for more on lek displays.*

Building the Best Nest

Q If robins make a nest in a bush close to the house one season, what are the chances they will go back to the same place the next season to build their nest?

A If the female survives the winter, and if she successfully reared young the first year, the chances are excellent that she will return to the same place the next year. If the nest failed for any reason, she will usually find a new place to build.

˅ ˅

Q Do all birds sleep in nests?

A No. Nests aren't designed to serve as beds for adult birds. They're really just incubators for eggs and hatchlings, and cribs for nestlings. Chicks of many birds, including ducks, geese, and relatives of chickens such as grouse, leave their nest within a few hours of hatching, never to return. Others remain in their nest until they can hop or fly.

Woodpeckers and other birds that nest in cavities often sleep in cavities even when they're not nesting. Cavities protect birds from excessive cold and heat and, even more important, from rain and snow and wind. But cavities aren't absolutely safe. Raccoons, cats, and some other predators can reach into cavities to pull out sleeping birds, which have no other way out. And cavities can foster parasites. Some birds roost in cavities only on very cold nights; otherwise, they roost on branches. As soon as young woodpeckers leave the nest, they move into abandoned cavities or excavate new ones for sleeping, and switch to new cavities fairly often.

This is why it's so important for birds to have both good nest sites and good roost sites in their habitats.

v v v v v v v v v v v v v v v v v v v v v v v v v v

Q **One day I looked out on my porch and a little bird was plucking fur from my golden retriever! Why was it harassing my dog?**

A The bird wasn't trying to bother your dog; it was just taking advantage of a convenient source of fur to gather soft insulation for its nest. Many birds, including Chipping Sparrows and Tufted Titmice, pluck fur from dead and sleeping animals. They also pluck tail hairs from horses. Once I saw a Tufted Titmouse pulling hairs from the tail of a raccoon. The tail was sticking out of a big cavity, and the raccoon was apparently sleeping inside. It hardly moved at all as the titmouse collected hairs; suddenly it rolled over, its tail slowly twisting around as the titmouse clung to it, finally flying off with a beakful of hairs.

The largest documented individual nest is that of a Bald Eagle in Florida; it was 20 feet (6 m) deep, almost 10 feet (3 m) wide, and weighed almost 3 tons. The largest nest on record built by social birds was a nest made by African social-weavers. The nest had 100 chambers. It was 27 feet (8.2 m) across and 6 feet (1.8 m) high.

NEST, SWEET NEST

When building a nest for the first time, most birds seem
to instinctively know what materials to use and how to
build it without having observed nest building before, perhaps from having been raised in a particular kind of nest,
but their skills improve with age and experience. Some
species, including turkeys, nighthawks, and Killdeer, make
a simple scrape in the ground, using their bellies to shape
it. Others, such as orioles, construct elaborately woven
nests using hundreds of strands of grasses or fibers. There
are hundreds of ways in which birds construct their nests!
Here are a few examples:

Peregrine Falcons nest on ledges, cliffs, and inside
special nest boxes with a gravel bottom set out to attract
them. The female doesn't bring any materials or make any
serious efforts at construction; she merely lowers her belly
and hollows out a little impression where she can lay her
eggs, to keep them from rolling off.

Woodpeckers build some of the most ecologically
important nests, because they provide roosting and nesting accommodations for many other animals after the
woodpeckers are done using them. A Pileated Woodpecker can hammer a hole about 2 inches (5 cm) deep
into hard wood with its bill. It can't hammer with any kind
of power deeper than that, because it can't align its body

to have leverage, so Pileated Woodpeckers seek out trees with healthy outer wood that are rotten on the inside. Male and female both hammer a round hole wide enough to pass through, and then reach in to grab mouthfuls of the rotted woody matter and spit them out to excavate a nesting chamber inside. During the nesting season, both parents incubate the eggs and then care for and feed the young throughout the daytime; the male stays inside the nest with the eggs or chicks at night.

Wood Ducks also depend on cavities but cannot excavate their own. Wood Ducks occasionally use old Pileated Woodpecker holes, but more often prefer holes formed when a rotten branch falls off a tree and heart rot creates a nest chamber. Wood Ducks don't collect nesting materials, but the female does line the bottom with down feathers she pulls from her breast and belly; they grow loose during the breeding season so this is easy for her to do.

Black-capped Chickadees can use an old Downy Woodpecker cavity or bird house; or they can excavate their own cavity in a rotten section of a birch, aspen, or other soft-wooded tree. Once the cavity is excavated, usually by both the male and female, the female builds the nest inside. She starts out using moss to form the bottom and sides, and then lines it with softer material such as rabbit fur.

(continued)

House Wrens nest inside old Downy Woodpecker or chickadee cavities, inside bird boxes, and inside all manner of other small spaces. Wrens have successfully nested inside old boots, overall pockets, truck axles, fish creels, and cow skulls. The male chooses as many potential nest cavities as he can find, covering the bottom of each with a platform of small sticks. The female chooses the one most to her liking and finishes building the actual nest in a depression in the stick platform; she lines this cup with soft materials.

Tree Swallows nest in old tree cavities or bird boxes. Females build the nest, which they line with feathers, mostly brought by the males.

Common Loons build their nests along a shoreline or on a mass of floating vegetation. Experienced pairs prefer floating masses or floating artificial nest platforms because they rise with water levels and aren't as readily flooded as a shoreline nest. Both adults work fairly equally on the nest, pulling or retrieving vegetation adjacent to the nest or from the lake bottom, and then tossing it sideways onto the nest. They both shape the nest by sitting and contouring the materials to their body shape. While sitting on the nest, they may pull materials from their surroundings to add to the nest. Early in the season, it usually takes more than a week for a pair of loons to build a nest, but later on it may take only a single day. This may be partly because they're more hormonally primed later on, partly because more vegetation is available further into the spring, and partly because late nestings may be replacement nests when the birds are in a bigger hurry to finish.

The Baltimore Oriole female builds her complicated hanging nest in three stages. First, she constructs the outer bowl of flexible plant, animal, or human-made fibers that can be very coarse — these provide the structural support. After she's put together the basic skeleton, she often works from inside the nest adding more fibers, especially as she moves onto the next stage. At that point, she begins weaving more springy, flexible fibers into an inner bowl, which fills in spaces and maintains the nest's shape. Finally, she adds downy fibers to line the nest. It usually takes her about a week to build the nest.

Barn Swallows work together to build their nest, though the female seems to take a leading role. Both birds collect mud in their beaks, often mixing it with grass stems to form rounded pellets, the basic "bricks." As soon as they form a new pellet, they fly to the nest site and attach it, first to a wall or other vertical structure, or to the top of a beam, eave, or other horizontal support.

(continued)

(Nest, Sweet Nest, *continued*)

Little by little they construct a narrow mud shelf, just big enough for them to sit on, and then build up the sides to form the inner bowl. If it's attached to a vertical wall, the nest is a semicircular half-cup shape. If it's built atop a beam or other bottom support, the nest will be more circular. The female spends a lot of time shaping it with her belly. Once the basic mud nest is constructed, they line it with grasses and then feathers. If things go well, they may complete the nest in three days, but sometimes it takes as long as two weeks, especially when the weather is unfavorable for gathering mud.

American Dippers, birds of gushing mountain streams in the West, nest on cliff faces or ledges, sometimes behind noisy waterfalls. Males may help with nest construction, but in some pairs the female does it all. They seem to prefer using wet materials for building. In 1908, D. Gale described the process for one female, who rejected her mate's "desultory contribution" to help. She built the nest "from the bottom up . . . the walls raised on all sides by forcing the building material into the wall from below. . . . (F)ibers on the outside . . . lie loose, plush-like, to lead the water from the dome roof as from a hay-cock. As this work on the walls dries, the insertion of other plugs . . . knits the whole densely. . . . by working from below, the lips or edges of the walls from all sides are closed up. . . . while working on the lower side the bird lay flat on the floor, spread her wings for a purchase and seemed to push with all her strength to ensure a well-caulked seam." All this took just one hour.

Bald Eagles usually begin building a nest together months before producing eggs, though in one case nest construction took only four days. Both mates gather fairly large sticks, though the female usually takes the lead in placing them. Grasses, mosses, and even cornstalks are used as fillers. Once the basic structure is in place, they'll add finer materials, such as Spanish moss, and then feathers. Pairs work on the nest throughout the nesting period and, especially, after their young have fledged each year and before the next nesting season. A single Bald Eagle nest may remain in use for several decades. As one bird dies, its mate will take a new mate, so what appears to be the same pair year after year may actually be different ones.

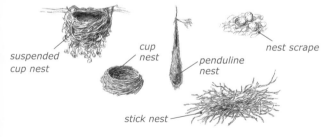

suspended
cup nest

cup
nest

pendiline
nest

nest scrape

stick nest

Q Last spring I watched prairie-chickens from a blind. Later in the morning when the chickens started leaving, suddenly Tree Swallows came in, flying back and forth just above where the prairie-chickens had been dancing. Do prairie-chickens have lice or were they stirring up bugs that the swallows were feeding on?

A The swallows were picking up little feathers that had gotten stuck in the prairie grasses while the prairie-chickens were preening, displaying, and fighting. Swallows line their nests with feathers, and after they've discovered that grassland grouse and prairie-chickens leave a lot of down feathers after their morning activities, the swallows make it a regular practice to check out the area every morning as the chickens leave.

∨ ∨

Q Are any martin houses guaranteed to attract Purple Martins?

A No. Purple Martins are declining over much of their range, making it difficult for people to attract them to a new nest site. You'll increase the likelihood of success by following recommendations from the Purple Martin Conservation Society. If you do set out a martin house, please make sure to evict any House Sparrows before they complete their nests. Do this by tossing out nest materials and closing the holes until martins appear. House Sparrows are a serious competitor with Purple Martins and are one of the major reasons for Purple Martin declines.

Purple Martin houses built as "apartment buildings" should not have ledges that connect one apartment to another. If a chick from one chamber wanders out and into the next apartment, its parents won't follow it there for feedings. Meanwhile, the parents nesting in that chamber may end up feeding the wanderer rather than their own nestlings if the wanderer is larger and appears hungrier. There are overall higher mortality rates in these kinds of martin boxes than in ones with dividers completely separating chambers, or in gourds.

ˇ ˇ

Q Help! We used to have a lot of Purple Martins, but they've been dwindling and this year there's just one pair left. House Sparrows took over the rest of the martin house, and all we ask is that they leave the one hole alone for the Purple Martins, but they're being so aggressive! I'm afraid we'll lose our Purple Martins for good. Is there anything we can do?

A House Sparrows are one of the worst problems facing Purple Martins. Having even one pair in a martin house is eventually going to eliminate the martins. Every year, the entry holes should be closed or blocked until martins return.

For the most current suggestions for dealing with House Sparrows in Purple Martin houses, visit the Purple Martin Conservation Association's website at *www.purplemartin.org*.

A "COMPOST" NEST THAT WARMS THE EGGS

Australian Brush-Turkeys have unusual nests that actually give off enough heat to incubate the eggs without the parents. The male uses his oversized feet to scratch together a pile of dead leaves, branches, and other fallen vegetation, constructing a huge nest, sometimes 12 feet (3.6 m) in diameter and 3 feet (1 m) deep. The nest looks like a messy pile of leaves but works as a compost pile to keep eggs warm. Female brush-turkeys mate with the males whose nests seem to provide the best incubation environment in which to lay their eggs. After the chicks hatch, they dig their way out of the nest and move on, ready to lead independent lives immediately.

It's illegal to collect nests — both state and federal laws protect birds and their feathers, eggs, and nests from being collected. Nests sometimes harbor mites, lice, and botfly larvae, so they probably aren't a good choice for keeping indoors anyway.

PROVIDING SAFE HOUSING FOR BIRDS

Unlike human housing, birdhouses don't have to meet any building codes and, sadly, many birdhouses on the market are made of metal or cheap plastic, both of which can get excessively hot or cold. And many houses have inappropriate hole sizes for the birds they are intended for. When a wren or chickadee house has a hole larger than 1⅛ inch (2.9 cm), House Sparrows can enter and kill the babies. When a bluebird box has a hole larger than 1⁹⁄₁₆ inch (3.9 cm), starlings can get in and kill them.

If the wood inside the box is too smooth below the entry hole, young bluebirds, ducklings, and other baby birds often cannot get the traction they need to get out of the box. Rough cut plywood is a good choice, or you can score the inside or add rough material below the hole.

Many people like the looks of a bird house with a little perch below the entrance hole. But birds accustomed to natural cavities and woodpecker holes don't need that perch, which is an open invitation to House Sparrows. If you buy a birdhouse that has a dowel or other perch below the entrance, break or cut it off.

Birds don't care about adorable designs. Keep designer birdhouses indoors, and set out birdhouses conforming to plans produced by authoritative sources such as the Cornell Lab of Ornithology *(www.nestwatch.org)*, state departments of natural resources or conservation, and organizations such as Audubon and the National Wildlife Federation.

Many birds never enter cavities or birdhouses — only a relatively small number of species do nest in birdhouses. The

(Providing Safe Housing for Birds, *continued*)

ones people are usually most interested in providing housing for include:

▶ American Kestrel (nest box)
▶ American Robin (nest platform)
▶ Barred Owl (nest box with very large entrance)
▶ Bluebirds (nest box)
▶ Chickadees (nest box or tube)
▶ Common Loon (nest platforms)
▶ Eastern Phoebe (nest platform)
▶ Flycatchers, Ash-throated and Great Crested (nest box)
▶ House Wrens (nest box or tube nest)
▶ Oak and Tufted titmice (nest box)
▶ Osprey (nest platform)
▶ Peregrine Falcon (gravel-covered ledge or open box)
▶ Prothonotary Warbler (nest box)
▶ Purple Martin (gourds or nest box with multiple compart-ments)
▶ Screech owls (nest box — same size as for Wood Ducks)
▶ Swallows, Tree and Violet-green (nest box — same size as for bluebirds)
▶ Wood Duck (nest box)
▶ Wrens (nest box or nest tube)

The Family Life of Birds

Q Birds seem so devoted to one another, courting and raising their young together. How many species mate for life?

A There is a great deal of variation among species, and even among individuals, with regard to mating systems. Swans, geese, and cranes may migrate long distances between nesting and wintering grounds, pairs remaining together throughout. Eagles and falcons may also migrate long distances and nest with the same partner year after year, but there is little or no evidence that they remain together except during the breeding season.

Florida Scrub-Jays are not the least bit migratory — birds spend their entire lives within a very small area — and pairs remain together as long as they both survive. Tropical wrens that remain on the same territory throughout the year not only mate for life but learn to sing complex duets, some so perfectly synchronized that it may sound like a single bird unless the listener stands between them.

There may be some randomness about other birds pairing for life. Chickadees tend to select mates from across the dominance hierarchy within their winter flock: the highest-ranking female mates with the highest-ranking male, the second-most dominant female with the second-most dominant male, and so on. Flocks are fairly stable from year to year, so some of these pairings may last several years. But when a high-ranking bird of one sex dies and the rankings shift upward, a shift in pairings usually occurs. Some birds that return to the same general nesting area year after year can end up with the same mate. With robins, this tends to occur more often when they were successful raising young the previous year. And for many species, there is little or no tendency to select the same mate in subsequent years.

Many species remain together for an entire season, but some only remain together for a single nesting and then both birds find new mates. After a pair of House Wrens fledges a brood, the female often finds a new mate while the male continues to rear the fledglings and attracts a new mate. In contrast, many ducks remain together during nest-building and egg-laying, but then males move on when the female starts incubating.

Some species don't remain together for long. Ruby-throated Hummingbirds don't form a pair bond at all. The male contributes to the survival of the young by driving away other birds and many large insects from nearby flowers, ensuring that the nectar supply won't be depleted when the larger, dominant female flies in to feed. This minimizes the amount of time she must spend away from eggs and chicks.

Male gallinaceous birds, such as turkeys and prairie-chickens, produce a display which attracts females. The most successful males may mate with many females. The females nest and care for the young without male assistance. In some cases, such as in turkeys, the males most likely to be selected as mates are also the most aggressive, and so females must try to keep their young ones away from these males until the young can protect themselves.

Whether birds of a pair stay together for a lifetime or for a nest period, there is also a great deal of variation in whether birds mate with birds other than their social partners. Ornithologists used to believe that the vast majority of bird pairs were faithful during a nesting, but DNA tests have revealed that more than one male has fathered many broods in more than 90 percent of all species tested. "Extra-pair paternity" is extremely common in ducks and some swallows, but extremely rare in Florida Scrub-Jays.

Teasing out the reasons why different species, sometimes closely related, have entirely different mating systems can keep amateur and professional ornithologists engaged for years.

⌄ ⌄

Q **Many species of birds share parenting duties after the eggs hatch. Do males help with the incubating process?**

A In some species yes, and in others no. Male and female Rose-breasted Grosbeaks share incubation duties. Some male Barn Swallows incubate a great deal and others don't, and only the males that do incubate develop a *brood patch* — a bare spot on the belly where the heat of the adult is transferred to the eggs. Loons take turns during the day but only the female incubates at night. Woodpeckers take turns during the day but only the male incubates at night.

In some species, such as hawks, owls, jays, and crows, the male hunts and brings food to the female, who remains on the nest virtually all the time, especially during bad weather. In some shorebirds, especially phalaropes (delicate birds that spin like tops as they swim to stir up food), only the male incubates and cares for the babies. In these cases, the female is the one who defends the territory.

There are almost as many reasons for differences among incubation strategies as there are species. Sandhill Crane females incubate while males guard. In the unrelated but super-ficially similar-looking Great Blue Heron, both sexes develop a brood patch and both take turns incubating.

Hummingbird and raptor females are larger than males, and in both groups the female incubates. But relative size cannot be the explanation because in geese and most blackbirds, the male is significantly larger than the female, yet again the female is the one who incubates. Red-winged Blackbird males are brilliantly colored; in this species only the cryptically colored female incubates. But both the dully colored female and the conspicuous male Northern Cardinal and Rose-breasted Grosbeak incubate, and both sexes occasionally even sing from the nest.

NOT AS EASY AS IT LOOKS

🐦 **Incubating takes a tremendous amount of energy,** especially in cold environments. A female Snowy Owl, for example, begins incubating the moment she lays her first egg, and if her mate is able to get enough food, she won't leave the nest at all for many days at a time. Although they look like they're not doing anything, incubating birds often lose weight during this period, especially if their mates are unable to bring them sufficient food.

Q There is a Wood Duck house on the river near my cabin, low enough that I can peek inside while I'm canoeing. Last year I watched the mother sitting on 12 eggs for weeks — she always sat tight whenever I looked in. But then one day she was gone and all the eggs were broken and looked like they'd been cleaned out. She never came back. What predator would have done this?

A It sounds like your ducklings all hatched out! Ducklings call inside their eggs to synchronize their hatching, and then when all the babies are dried off, they jump from the nest and follow their mother to water, never to return unless one of them, as an adult, chooses that box to nest in. If a predator had found them, you'd have found signs of blood and at least a few broken but unemptied eggs.

˅ ˅

Q A female cardinal nesting in our yard died by flying into a window. What will happen to the eggs? Will her mate take care of them?

A Male cardinals can incubate eggs, although the female typically incubates much more. This male may take over entirely and raise the young on his own. Some cardinals in this situation, however, will abandon the nest and move on to find a new mate with which to start over. This approach can actually allow a bird to raise more young that summer. If the male tries to hatch out and raise the original nest of eggs, he's almost certain to lose at least some of the babies in the nest because it's hard for a single bird to find enough food for four or five nestlings. If the cardinal does abandon the nest, a chipmunk, jay, crow, or squirrel may come in and eat the eggs. It's heartbreaking, but somehow birds muddle through.

KEEPING AN EYE ON THINGS

If birds are nesting near you in an accessible spot, it's hard to resist checking on the eggs and the chicks. It's okay to watch discreetly, but too much attention could cause the parents to abandon the nest. It's best to peek in on nests in midafternoon (when parents are feeding young the least often), for a very brief time, and not every single day.

If you enjoy finding and watching bird nests, consider joining the NestWatch citizen-science project, coordinated by the Cornell Lab of Ornithology and Smithsonian Migratory Bird Center. Visit *www.nestwatch.org* to find out how to watch nesting birds safely and to record your observations. The information you gather is valuable to scientists studying nesting birds and the factors that lead to their failures or successes.

Q Why don't I ever see baby pigeons?

A Pigeons make their loosely constructed cup nests in crevices and on ledges. They lay two eggs, which hatch after 18 days, and the babies, called squabs, tend to sit tight for more than two weeks. By day 18 after hatching, they can walk and may wander a bit along their ledge, but they don't open their wings and fly down from it until they're completely feathered and able to fly, sometime between day 25 and 32. By this time, they are as large as adults, but they're still dependent on their parents for protection and food.

⌄ ⌄

Q Last summer after a storm we found a tiny baby robin that had fallen from its nest. I put it back in the nest, but my neighbor told me the parents would reject it if it smelled like human hands. I'm not sure, but it looked like the parents were still feeding it after I put it back. Can birds really smell our hands?

A Probably not, and they don't reject young birds after people touch them. Most birds have a sense of smell that is as poor as ours, or poorer, and they identify their babies the same ways we do — by how the babies look and sound. If you find a nestling — a baby bird that can't yet hop, walk, or fly — and if you know where the nest is, the best thing you can do is return the little bird to the nest.

Q I had a Mourning Dove nest in my hanging basket. I couldn't resist the temptation and tried to feed the nestlings some millet, but they got scared and flew away. Will they ever come back to the nest? Can their parents find them? I am upset that I came that close to them. They did have full feathers, and I believe that they were about two weeks old. It seemed like the parents were staying away for longer periods of time. This morning I heard the cooing of an adult near my house and felt sad — was that the parent looking for its young?

A Don't feel bad. The startle response that causes nestlings to flee when they recognize danger doesn't kick in until they are mature enough to survive. Their parents will search for them and will almost definitely find them. Mourning Doves usually fledge between 12 and 14 days, so these may well have been ready to go anyway.

The young will never return to the nest — they'll be roosting on branches now, exactly as they would if they had fledged on their own without being startled. The parents may or may not reuse this plant for a nest site next time — we never know what will lead them to make that decision. The adult singing is part of getting revved up for the next breeding period, because they will renest after these babies are no longer dependent.

While birds are still nestlings, they usually have a special diet, so "help" from people isn't generally helpful. Mourning Doves feed their babies "pigeon milk," which is surprisingly similar to mammalian milk, though it's manufactured in a part of the bird's digestive system called the "crop." As baby doves grow, the parents' bodies start mixing the pigeon milk with a slurry of regurgitated seeds, so when the babies do start feeding on their own, they'll be able to digest seeds.

∨ ∨

Q **Someone told me it's illegal to take care of baby birds. Is this true?**
A Yes. This may sound cruel when we find a desperate little bird barely clinging to life. But in this event, enforcement is usually lenient if you do your best to keep the bird alive while you find a licensed wildlife rehabilitator to take the bird. Native birds are protected by state and local laws, especially the Migratory Bird Treaty Act, and cannot legally be kept as pets.

Every species of baby bird has specific dietary and care needs, and it's very easy to accidentally injure them. And even if we can keep them alive, it's critical to a bird's survival and quality of life to be educated by its natural parents or surrogates. Sometimes when young birds leave the nest prematurely and cannot yet fly, people try to raise them. But if these birds are well-feathered and can hop, they quickly become skilled at hiding, and their parents can usually find them and continue to feed them raise them.

Rehabilitation facilities usually know what to do to provide at least some of the skills wild birds need to survive and thrive, but baby birds have a much better prognosis for a long, natural life if they are raised by their parents.

"I DO IT!"

🐦 **Individual baby birds,** even in the same family, may have very different "personalities." One day I spent hours watching two young female Pileated Woodpeckers learning to feed with their parents. One would try half-heartedly to probe at a dead tree as her father pointed out the best spots, but she'd quickly start to whine and flap her wings, begging for him to feed her. Meanwhile, the other daughter was with the mother at a different stump that looked no richer in bugs. This daughter probed busily, sometimes pulling out a big grub and toying with it before eating it. The mother offered her a couple of grubs, but this youngster just ignored her, reminding me of trying to dress my daughter when she was a toddler — she'd refuse my help, insisting, "NO! Katie do!"

Q Why do birds leave the nest before they can fly?

A It's to any young bird's advantage to leave the nest as soon as it safely can. People tend to think of nests as safe, cozy little homes, but predators have an easy time locating and raiding a nest of calling baby birds, and the warmth and high humidity of nests make them breeding grounds for dangerous avian parasites.

Some species don't remain in the nest more than a matter of hours. These babies follow one or both of their parents, learning what to eat and how to avoid danger and where to hide while resting or when in danger. But being flightless does put ducklings, Killdeer chicks, and other "precocial chicks" at risk from all kinds of predators, so most of the birds that raise young this way must produce a lot of eggs in a single clutch to have a reasonable expectation that over their lifetime, at least two babies will survive to replace them.

Baby birds that remain in nests and have a fairly long helpless stage cost their parents' significant time and energy. Parent hawks, herons, songbirds, and other types of birds work from sunrise to sunset every day to get their young fed and out of the nest as quickly as possible. After fledging, the young birds are more spread out, and the parents can lead them to different spots every night, enhancing each one's chances of survival.

Some species, such as swallows, woodpeckers, and other cavity nesters, nest where there are no nearby branches for young to grab onto when they first leave the nest. Unless startled by a predator, young of these species tend to remain in the nest until they are strong fliers.

^v^v^v^v^v^v^v^v^v^v^v^v^v^v^v^v^v^v^v^v^v^v^v^v^v^v^v^v^v^v^v^v

All About Eggs

Q How many eggs do birds lay?

A This depends on the species. Some species, such as albatrosses and some penguins, lay just one egg each time they attempt to raise young. Despite their difference in body

size, Ruby-throated Hummingbirds and Common Loons lay
two. Many songbirds lay four
or five eggs, but chickadees and
House Wrens usually lay six to
nine and the even tinier kinglets
may lay even more! Ducks and
grouse can lay a dozen or more
in a single clutch. If something
destroys the nest early in the season, most birds can start over,
but if it's too late, they give up for the year.

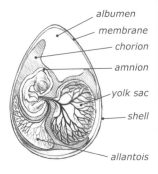

albumen
membrane
chorion
amnion
yolk sac
shell
allantois

Many birds are *indeterminate layers*, meaning if their eggs are
removed one by one, they'll continue laying more for a long
time. People take advantage of this with domestic poultry, but
indeterminate laying is also present in wild birds. In one experiment, an egg was removed each day from a Northern Flicker
nest (a species that normally lays 6 to 9 eggs in a clutch), and
the female produced a total of 71 eggs in 73 days.

∨ ∨

Q **How long does it take for eggs to hatch?**

A It depends on the species. Small songbird eggs can hatch
in just 11 days, Ruby-throated Hummingbirds in 12 to
14 days, Common Loons and many ducks in about 28 days,
Bald Eagles in about 35 days, and Emperor Penguins in about
65 days.

On average, clutch sizes are smaller in the tropics, where adults are long-lived, than in the temperate zone, where birds live fast and die young.

Q What part of the egg turns into the chick?

A The yolk is a single giant cell that divides to form the chick. Of course, the whole yolk doesn't split in half — the cells that divide from it are very tiny and grow to surround the yolk as it shrinks, providing the nourishment for the growing chick.

❤ ❤

Q My Cockatiel lays eggs all the time, but there isn't a male in sight. I thought you needed two parents to make a baby.

A You do need two parents to make a baby bird. But during the breeding season (which for Cockatiels is much of the year), females ovulate whether or not they mate. Unfertilized eggs, like store-bought chicken eggs, have a yolk and albumen but can't produce a chick.

It may be a good idea to let her build a nest and lay the eggs in there. Once she has a full clutch, she'll start incubating them and will stop ovulating. If you remove the egg every day, she'll keep laying eggs. Producing eggs requires calcium, so if she's continually laying eggs, she may suffer from calcium deficiency.

Q While at our family cabin in Michigan, I found a Cooper's Hawk egg on the ground, about a hundred yards from where the nest was. It was cold to the touch, but I picked it up and kept it warm, just in case. I'd like to know if there's a way to find out if the chick inside is still alive.

A To see the chick inside, people candle eggs; that is, they hold them up so they're backlit by a bright light. But please don't try to hatch this egg. If it was dropped, it's very likely the embryo was killed. It's even likely that the parents tossed it out when it didn't hatch after other nestlings did hatch. If something else happened and a live egg was dropped by a predator or fell out of the nest in a storm or accident, its chances of hatching into a healthy chick are low. Being shaken, the impact of falling, and being significantly cooled after the onset of incubation could each be problematic.

LET SLEEPING BIRDS LIE

Water birds sometimes sleep in the water. Some sleep on tree branches or in cavities, too. Some birds can be literally half asleep — they close one eye and allow one half of their brain to sleep while the other eye and half of the brain are engaged in watching for predators.

Most songbirds find a secluded branch or a tree cavity, fluff out their down feathers beneath their outer feathers, turn their head to face backward, tuck their beak into their back feathers, and close their eyes.

PART THREE

All About Birds, Inside and Out

CHAPTER 9

The Inside Story: How Birds Work

When we sit down to a Thanksgiving turkey dinner, our minds are usually filled with thoughts of family and our many blessings, but once in a while, we can't help but wonder why turkeys have dark meat and white meat while geese have only dark. What's the gizzard all about? And why don't we ever see a turkey's lungs?

Whether we're questioning our food or trying to understand how vultures can breathe at high altitudes or how a gannet can plunge into water at 60 miles per hour, bird insides are fascinating, leading to all kinds of questions.

∧∨

From the Inside Out

Q One time when I was on a cruise, the ship naturalist pointed out a gannet, flying up high and then dropping down like a bullet to dive into the water. How does it hit the water at such high speed without injuring itself?

A Gannets and their close relatives, the boobies, can plunge straight into the water from as high as 120 feet, hitting the surface at speeds that can exceed 60 miles per hour. To survive the impact, they dive so that their pointed bills enter the water first, followed by their streamlined bodies. Gannets have unusually strong skulls and rib cages, and their respiratory system includes tiny air sacs between their skin and their muscles that absorb the shock when they hit the water. Their

external nostrils are closed, and their "secondary
nostrils," next to the bill, are covered by move-
flaps that are closed when they hit the water.
They can see schools of fish from the air,
track them as they dive, and follow them
underwater — a semitransparent inner
eyelid, the "nictitating membrane,"
protects their eyes.

˅ ˅

Q When I was visiting my grandfather's farm, he
butchered a turkey for dinner and let me look at the
organs. I could figure out where the heart, stomach, liver,
and intestines were, but I couldn't find the lungs at all!
Where were they?

A Birds have high metabolic rates, and they exert intense
energy while flying long distances, which requires pro-
portionally more oxygen than mammals use. Mammalian
lungs are huge, yet very light, and positioned right where a
bird's center of gravity must be for efficient flight. Birds have
an extremely specialized respiratory system, unique in the animal kingdom, that allows them to extract much more oxygen
from the air they breathe in than we can. To accomplish this,
avian lungs are very small and flat, and rigid, not expandable.
That turkey's lungs were there, fitted perfectly against the bird's
back ribs, but so flat that they didn't look at all the way you
expected.

When birds breathe in, the tide of air that flows through the trachea is split. A small portion of it goes straight into the lungs for immediate exchange of oxygen and carbon dioxide. From there it flows into air sacs in the forward part of the body. Like bellows, when the bird exhales, these anterior air sacs push the oxygen-depleted air back into the trachea and out through the bird's nostrils or mouth. A larger part of the inhaled air goes from the trachea straight back to posterior air sacs, and when the bird exhales, these sacs send the fresh air through the lungs and out through the trachea.

So all the air in the lungs at any given moment is fresh, unlike the air that remains and becomes mixed with carbon dioxide in the alveoli of mammalian lungs. The air sacs are huge in living birds, but most of the space they fill is behind the relatively heavier organs in the bird's center of gravity.

When birds die, the weight of their organs and skin, along with air pressure from outside the bird, instantly flattens most of the air sacs, so you wouldn't have been able to easily detect them in the turkey unless you put a straw down its trachea and blew — then the air would flow through the lungs to inflate the air sacs, which look like clear balloons.

˅ ˅

Q When I saw the IMAX movie *Everest*, I noticed big black birds like crows up near the summit, where virtually all the people needed oxygen tanks. How can birds breathe at such high altitudes?

A The trick isn't in breathing at high altitudes; it's in actually getting oxygen from those breaths into the blood and getting carbon dioxide out when there is so little oxygen in the atmosphere. Bird lungs, a crisscrossing matrix of tiny air tubes and parallel capillaries, are exquisitely designed for this. Air passes through the tiny tubes in a direction countercurrent to the blood flowing through capillaries, making it extremely easy for the blood to pick up oxygen from the air tubes as the tubes pick up carbon dioxide from the blood to exhale. Every air tube in avian lungs is interconnected; bird lungs have no dead-air spaces such as the alveoli, where gas exchange takes place in mammalian lungs.

The whole point of the respiratory system is to get the right amount of oxygen to all the cells, which happens in conjunction with the circulatory system. Strong flyers tend to have smaller red blood cells than flightless birds and weak flyers. This difference is important because the smaller a cell is, the relatively larger its surface area for gas exchange.

As with human athletes training at high elevations to build up their red blood cells, birds that move to higher elevations or are subjected in laboratories to lower air pressure produce more red blood cells. Birds have slightly less hemoglobin in their red blood cells than do mammals, but avian hemoglobin is more efficient at picking up oxygen.

Those birds, by the way, are Alpine Choughs.

HEARTBEATS AND BREATHING RATES

Because birds inhale and exhale relatively large amounts of air in each breath, as compared with mammals, their breathing rates are much slower than those of mammals of comparable size. The smallest hummingbirds take about 250 breaths per minute while the smallest shrews take about 800. Of course, flying birds breathe more rapidly than resting ones: a duck at rest takes about 14 breaths per minute while a flying duck takes about 96.

The rapid rate at which oxygen is taken up in the blood is reflected in a bird's rapid heart rate. A mammal's average heart rate is about 3 times its respiration rate whereas a bird's average heart rate is over 7 times its respiration rate. A chickadee's heart beats about 500 times per minute at rest and doubles that during activity. To pump blood efficiently, the heart of a bird is larger than that of a mammal of comparable size, and the heart is relatively larger in smaller birds than in larger ones. It's also larger in alpine birds than in their relatives who live at lower elevations.

Q **Why do chickens and turkeys have white breast meat while ducks and geese have all dark meat?**

A Birds and mammals have two kinds of muscle fibers: white, or "fast twitch," fibers and red, or "slow twitch," fibers. Red fibers may not twitch as fast as white ones, but the red fiber cells are richly supplied with mitochondria and red muscle tissues are heavily oxygenated with a rich blood supply. This enables these fibers to work steadily for long periods, so birds capable of long bouts of walking or swimming, such as turkeys, chickens, ducks, and geese, have mostly red fibers in their leg muscles. Birds capable of sustained flight, such as ducks, geese, songbirds, hummingbirds, and the like, have mostly or even all red muscle fibers in their wing and breast muscles. Most birds have more red than white muscle fibers; some hummingbird and House Sparrow muscles are made of 100 percent red fibers.

White meat comprises mostly fast twitch white fibers but usually has at least a few red fibers, too. Gallinaceous birds in general (including turkeys and grouse) have white meat in their pectoral muscles, which means the muscle fibers are mostly the fast twitch variety allowing the birds to take off in flight with a powerful burst. These muscles are seldom capable of sustained flight, though wild gallinaceous birds such as turkeys and grouse can sometimes fly a mile or more before tiring.

The red fibers in "dark meat" are more useful for more purposes because they're capable of sustained activity, but the trade-off is that they require more nutrients and a rich supply of blood vessels. Since gallinaceous birds don't normally fly long distances anyway, they're better off with the lighter, lower-maintenance white muscle fibers, giving them white meat. Ducks and geese, which need to engage in both sustained swimming and sustained flying, don't have white meat at all.

Domesticated chickens and turkeys have been bred to have a higher percentage of white fibers in their white meat, as well as to have more meat in the first place. So Wild Turkeys may have darker colored meat than do domesticated turkeys, but it's still considered white meat.

ᵛ ᵛ

Q I'm pretty good at imitating bird calls, but there are some songs, such as the Wood Thrush's, that I find impossible to whistle. I can't come close to imitating the shimmery, complex quality of its song. How does it do that?

A Birds produce sounds with their *syrinx* (literally "song box"), which is situated at the very base of the trachea and the top of the two bronchial tubes. By controlling the muscles of the syrinx, some birds can make one set of sounds with the left branch and a different set of sounds with the right, producing harmony with their own voice. The most complicated sets of muscles in the syrinx belong to the thrushes, which is why so many of their songs sound so ethereal.

ᵛ ᵛ

Q I was amazed to read that loons spend the winter on the ocean. How can they survive — are they able to drink saltwater?

A Yes, loons can drink saltwater when they are at sea because they have special glands that remove salt from their bloodstream. When they hatch, loons are freshwater birds, eating freshwater fish and drinking fresh water — until

"A POEM IN BONE"

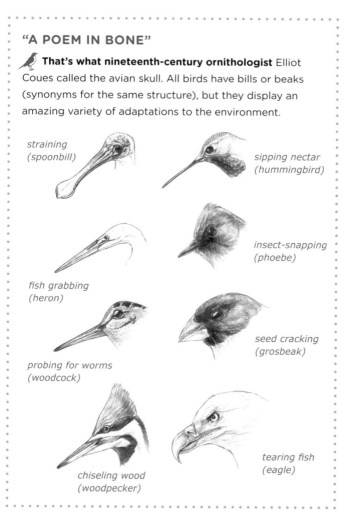

That's what nineteenth-century ornithologist Elliot Coues called the avian skull. All birds have bills or beaks (synonyms for the same structure), but they display an amazing variety of adaptations to the environment.

straining (spoonbill)

sipping nectar (hummingbird)

fish grabbing (heron)

insect-snapping (phoebe)

probing for worms (woodcock)

seed cracking (grosbeak)

chiseling wood (woodpecker)

tearing fish (eagle)

their first migration. The moment a loon gets a taste of salt-water, two glands near its eyes swell up and start removing excess salt from the bloodstream. The salt is excreted in the form of thick tears or mucous via the nasal passages. Our own bodies excrete excess salt by way of our kidneys and sweat glands, and our tears are as salty as our body tissues. The tears of loons in winter are much saltier than ours, or than the loons' body tissues, because the excess salt is concentrated before excreting. When loons return to freshwater in spring, their salt glands shrink for the season. If a loon were to cry in summer, its tears would be no more salty than ours.

Some ocean birds, such as albatrosses, have huge salt glands that excrete the salt through exceptionally long nostrils, which give albatrosses, petrels, and shearwaters the name "tubenoses." Excess salt can dribble out, as it does for loons, or it can be forcibly blown out through these specialized nostrils.

Some land birds, such as Budgerigars (our familiar pet "parakeets" or "budgies"), also have large salt glands. Wild Budgerigars, native to the interior deserts of Australia, drink from waterholes filled with brine water.

Digestive Issues

Q I once found a baby robin hopping on my lawn. There was a cat around so I picked the robin up and put it in a shrub where it would be safer. While I was carrying it, it opened its beak, begging for food. I was surprised that the inside of its mouth was bright yellow. Even weirder, the roof of the mouth had a most bizarre pattern, with something that looked like a serrated diamond pattern in the center. What was that?

A Many young birds, including robins, have mouths that are brightly colored on the inside. The color helps stimulate their parents to feed them when they open their mouths. On the roof of the mouth, you saw the opening to the bird's nasal passages and its Eustachian tubes. When birds breathe through their nostrils, the air flows through the nasal passages to the *glottis*, a fairly large opening on the bottom of the bird's mouth where the trachea begins.

v v v v v v v v v v v v v v v v v v v v v v v v v v v

Q Birds don't have teeth, do they? How do they chew their food?

A You're right, birds don't have teeth. They use their bills to handle and prep food before they eat it, but they don't actually need to chew their food before swallowing it. Hawks use their sharp bills to rip apart prey, then swallow chunks of meat. Finches and sparrows use their bills as a nutcracker to open hard-shelled nuts or seeds, which they then swallow whole. Birds that eat seeds can have many small mucous glands

289

inside their mouth to help them lubricate the seeds before swallowing, and salivary glands that secrete enzymes to begin the digestive process.

We humans have a gag reflex that usually prevents us from swallowing large items until they've been pulverized by our teeth. But even fairly small birds can swallow chunks so big they'd make us choke. A tiny Saw-whet Owl has no trouble downing a whole deer mouse, and a Great Blue Heron can swallow huge fish whole. Birds that eat small food particles, such as insect eaters, may have a fairly narrow esophagus, but those that eat big chunks have an extensible esophagus that stretches to accommodate the food; the skin of their neck is equally stretchy.

⌄ ⌄

Q I understand that "bird's nest soup" is a delicacy in Chinese cuisine and that the nests are made of saliva. Is that true?

A Yes, the nests used in soup are constructed by Asian swiftlets whose gluey saliva hardens into a cup-shaped nest. The most heavily harvested nests are from the Edible-nest Swiftlet and the Black-nest Swiftlet, found mostly around the

Many sea birds swallow fish so large that several hours after swallowing one, while the fish head is slowly digesting in the bird's stomach, the tail may still be sticking out the mouth.

coasts of southern China and Southeast Asia, and inland where suitable caves exist. It takes about 35 days for each male to build a nest, constructing a shallow, gluey cup out of strands of saliva containing predigested seaweed, stuck to the wall of a cave. These nests contain high levels of calcium, iron, potassium, and magnesium.

Collecting them for soup is both difficult and dangerous. Nest collectors climb flexible rattan ladders dangling from the ceiling of the caves, then inch along bamboo ladders to gather nests adhering to the roof, as much as 200 feet above the hard cave floor.

Nests are collected twice a year. The first collection takes place after the swiftlets have built the nests but before laying eggs. The males rebuild, and at this point pairs are allowed to raise their young. After the nestlings fledge and families leave the cave, these second nests are collected. Swiftlets cannot be raised in captivity and their nests are fairly small, making the nests extremely valuable commercially. One pound of "white" nests (produced by the Edible-nest Swiftlet) can cost almost $1,000, and one pound of "red blood" nests (which may get their color from the insects the Edible-nest Swiftlet eats) can cost more than $4,000! Black-nest Swiftlet nests aren't as valuable: they get their color because the bird mixes feathers with the saliva, and the feathers must be removed before using the nests for soup.

Three of the four North American swifts also use sticky saliva to hold their nests together and attach them to the walls of chimneys, hollow trees, and other structures, but their nests are hardly edible, being constructed of twigs, pine needles, weeds, and so forth. The Black Swift uses no saliva at all; its nest is just a shallow cup shaped out of mud or mosses on seacoast cliffs or behind mountain waterfalls.

Gray Jays have another unusual use for saliva: They coat chunks of food with copious amounts of saliva before fastening them to trees for storage. Researchers speculate that the saliva may help hold the food in place and possibly even retard spoilage somewhat.

v v v v v v v v v v v v v v v v v v v v v v v v v v

Q **Why do domestic birds such as chickens need grit with their food? Do wild birds also need grit to help them digest?**

A Birds, like mammals, can't absorb any nutrients until their food has been liquefied, which in the case of birds takes place almost entirely in the stomach. A bird's stomach has two chambers. Food first reaches the *proventriculus*, a glandular chamber where powerful acids start dissolving the food. Then the food passes into the *gizzard*, a muscular chamber that mashes the food.

> If you ever have the opportunity to get your hands on a chicken that has recently been feeding, hold its stomach up to your ear to hear the grinding sounds.

The gizzard is extremely powerful in birds that eat large or thick-walled seeds, such as turkeys, ducks, swans, chickens, sparrows, and finches. Seeds must be thoroughly pulverized to be digested. To accomplish this, most seed-eating birds pick up and swallow grit that remains in the gizzard for long periods as it is slowly dissolved and ground up by the digestive process.

Some kinds of grit have valuable minerals. For example, the slowly dissolving pieces of a pet bird's cuttlefish bone release calcium into the bird's bloodstream while it grinds up seeds in the bird's gizzard. In the same way, many wild birds, from loons and swans to sparrows and finches, swallow grit to aid in digestion and obtain necessary minerals.

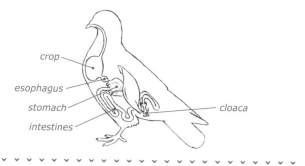

crop

esophagus

stomach

intestines

cloaca

✓ ✓

Q **I've heard that a major problem facing many birds is lead poisoning. Why is that?**

A Most of the pieces of grit that birds pick up are rich in calcium and other minerals, providing an important nutritional component as well as a digestive aid. Tragically, sometimes birds pick up bullets, tiny lead pellets from shotguns, or lead sinkers broken off from fishing lines. As these are gradually dissolved in the gizzard, the lead enters the bird's bloodstream, making the bird weak and disoriented, and often killing it. Waterfowl were succumbing to lead poisoning in such significant numbers that lead shot has been banned for waterfowl hunting in the United States since 1991. But lead shot is still permitted in most areas for hunting upland game birds such as grouse and pheasants, and hunters are permitted to use lead bullets for deer and other large game.

Scavengers, such as vultures, condors, and eagles, that feed on pheasants and other upland game birds that have been shot but not retrieved are often poisoned by lead shot. Those that dine on large chunks of meat or gut piles from deer may ingest fragments of lead bullets that, like lead shot, remain in the gizzard, dissolving and slowly poisoning the birds.

The most common cause of mortality for California Condors that have been released back into the wild has been from lead poisoning. Chemical analyses of the lead in these birds matches the forms of lead used in the manufacture of bullets, so in 2007, California passed a law requiring deer hunters to use copper bullets in specified areas where condors forage.

Many scientists and conservationists believe lead bullets and shot should be banned for all hunting, arguing that other nonendangered species are also ingesting lead. Their case is gaining supporters among human health advocates as new research is finding significant lead amounts in venison eaten by hunters and donated to food shelves.

Q Why do owls regurgitate pellets?

A After an owl's prey is dissolved in the glandular stomach, the gizzard squeezes the digestible material into the intestines while the bones, fur, teeth, and other indigestible matter remain. When all the liquid is finally squeezed out, what remains is spit out as a pellet. Hawks, crows, nighthawks, gulls, and some other birds also occasionally or even daily spit out undigested material as pellets, but owls produce the most solid ones.

The pellets collect beneath branches where owls roost during the day, providing both a useful clue for birders as to where an owl might be spotted and important information for researchers who are surveying small mammal populations. Such surveys help to determine how best to deal with agricultural pests that cause major economic problems for farmers.

Because owls usually swallow small prey whole or in two bites, the bones in an owl pellet tend to be fully intact and are usually easy to identify. Hawks, however, typically tear their meat apart, eating fewer bones in the first place, and hawk stomach secretions are significantly more acidic than those of owls, dissolving many of the bones they eat.

A bird's skull must house its enormous eyes and a brain large enough to coordinate keen vision with precise muscle control. Because their food is pulverized in the gizzard, birds have neither teeth nor a bony jaw to support them.

Q How do bird intestines compare with those of humans? Are they longer or shorter relative to their size?

A Birds that eat nectar, soft insects, or meat usually have very short intestines, but birds that digest grasses and other foliage or grain tend to have significantly longer ones. The small intestine of an Ostrich, a bird that grazes on grasses and seeds, is 46 feet (14 meters) long, or double the length of a human small intestine!

Of course, flying birds can't carry much weight, and so to reduce the weight they must lug around, most food is digested very quickly and efficiently. Rather than keeping food in the intestines for a long time, birds have a relatively huge liver (the heaviest internal organ) and pancreas, both situated in the optimal place for a flying creature's center of gravity. Both of these organs secrete juices that speed up the digestive process. What remains in the intestines after all the nourishment has been absorbed enters the bird's cloaca to be ejected when the bird poops.

The seeds of berries eaten by young Cedar Waxwings appear in their droppings just 16 minutes later, and seeds of elderberries eaten by thrushes pass through the digestive system in a mere 30 minutes. A Northern Shrike can digest a mouse in 3 hours, while the "bearded vulture" or Lammergeier of the Mediterranean (related to North American hawks) can completely digest a cow vertebra in a day or two.

Q Why is bird poop white?

A You may not care to examine bird droppings too closely, but if you do, you'll see they're usually made up of two different components. The brownish or dark greenish part is the fecal matter, or the poop. This is what remains after food has gone through the entire digestive system. The white part is actually urine.

Our urinary system filters our blood of impurities, which build up even before birth. Mammals eliminate these wastes as urea, a clear, yellowish fluid that is highly toxic and must be diluted with huge quantities of water. Birds and other animals that hatch from eggs eliminate the wastes from their kidneys as uric acid, a substance that doesn't need to be diluted in huge quantities of water because when concentrated it precipitates into a chalky, whitish substance.

ˇ ˇ

Q Do birds poop in flight?

A Birds tend to poop right before or during takeoff, and it takes a few minutes to build up another supply, so on short flights, no, they don't. But on flights that last longer than a few minutes, yes, they can and do poop in the air. Normally if they're fairly high up and going fairly fast, the poop atomizes long before it reaches the ground.

WATCH OUT FOR BIRD POOP

A few birds that drink huge quantities of water, such as hummingbirds, can produce urea as adults. Hummingbird droppings are usually droplets of clear liquid with tiny particles of fecal matter suspended within. Captive hummingbirds fed nothing but sugar water can't survive for long, but while they do, they produce urine with little or no poop.

Seabirds, cormorants, gulls, and other birds that take in a lot of water with their fishy diets don't produce urea, but they do release large amounts of uric acid compared to fecal matter. Those big white splats can be valuable; where these birds collect in large numbers, this *guano* can be collected for nitrogen- and phosphorus-rich fertilizers. Geese eating huge amounts of grasses produce smaller quantities of urine than fecal matter (which is mostly the indigestible cell walls of those grasses, hence its color and consistency).

The Green Heron, Black-crowned Night-Heron, and American Bittern have long been nicknamed "shite-pokes." Apparently when the term was first used in the 1770s, the word "shite" meant the exact same thing as a more common term used today, but it was a perfectly respectable way of referring to these birds' habit of shooting out a noticeable stream of poop on takeoff.

Facts about Feet

Q Every time I've ever seen an owl, in real life or in a photo, it had only two front toes. But last week I found a dead owl on the road, and it had three front toes. Was it a mutant?

A No. All owls have three toes that face forward and one that faces backward. You only see two toes in front in perched owls because one of the forward toes is opposable, like our thumb, and rotates to the rear when the owl sits on a branch.

Another bird of prey shares this feature — the Osprey. Osprey feet are specialized for catching fish, with characteristic papilla called *spicules* covering the bottom of their feet. When they grasp a fish, the two always-facing-forward front toes are balanced by the rear toe and the opposable toe, making it harder for a thrashing fish to escape.

Bald Eagle feet lack some of the specializations of Osprey feet. When an eagle catches a fish, its three forward toes from each foot grasp one side of the fish while the single back toe from each foot grasps the other side. Eagle talons are powerfully strong with long, sharp claws, but if a fish thrashes hard enough, it may pull out of the eagle's outer toes, forcing the eagle to shift its grip. Sometimes at this point the eagle may drop the fish. Eagles feed also on carrion and have a habit of stealing fish from Osprey, so despite being a bit less specialized, they have a wider margin of error.

I often have the fun of speaking to classrooms about birds. Once after I explained about eagles sometimes dropping their fish, a fifth-grade boy raised his hand and told me about a walleye he had caught that summer. "The fish had six scars on one side of its body, and two scars on the other, and I told my mom they looked like they came from eagle claws. But she said I had a big imagination." He was elated when I told him he'd make a great forensic detective.

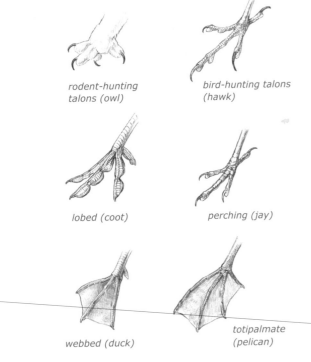

rodent-hunting talons (owl)

bird-hunting talons (hawk)

lobed (coot)

perching (jay)

webbed (duck)

totipalmate (pelican)

On March 30, 1987, an Alaskan Airlines jet was forced to make an unscheduled stop after the plane had a midair collision with a fish. During takeoff, right after it was too late to stop, the pilot suddenly saw several eagles flying above the plane. He gave a sigh of relief as he realized his flight path was safely beneath the massive birds, but one eagle apparently was startled and dropped its fish, which slammed into a small window at the top of the cockpit. The passengers, crew, and eagle all survived without injury, and damage to the plane was minimal. The only fatality was the fish.

Q Great Blue Herons look a lot like Sandhill Cranes, but someone told me they're not related, and that it had to do with their feet. True or false?

A Herons and cranes do look a lot alike, superficially; but they have evolved in entirely different lines, and their feet show some important differences. Herons nest and roost in trees, and so they have a long, strong back toe that allows them to cling to branches. Cranes nest on the ground and never perch in trees. Their back toe is tiny and raised — completely worthless for perching. If you see a footprint with long, unwebbed toes, that was made in sand or mud, you can tell if a heron or a crane made it by looking for signs of a back toe.

LONG LIVE THE BIRDS

Most wild birds never see their first birthday. There are so many dangers out there that birds produce fairly large numbers of young simply to ensure that during their lifetime there will be offspring to replace them.

But once an individual bird has survived its first year and has developed skills to negotiate each season, its life expectancy goes up dramatically. We can tell how long some individual birds have lived in the wild if they were banded with a U.S. Fish and Wildlife Service leg band and were recaptured or found dead at a later time. The ten longest-lived species recorded by the U.S. Geological Survey's Bird Banding Laboratory in Laurel, Maryland, as of February 2008 are:

▶ Laysan Albatross   50 years 8 months

▶ Black-footed Albatross   40 years 8 months

▶ Great Frigatebird   38 years 2 months

▶ White Tern   35 years 11 months

▶ Sooty Tern   35 years 10 months

▶ Wandering Albatross   34 years 7 months

▶ Arctic Tern   34 years

▶ Red-tailed Tropicbird   32 years 8 months

▶ Black-browed Albatross   32 years 5 months

▶ Atlantic Puffin   31 years

Every one of those "top ten" birds is a species that spends its life on the ocean. But some inland species have surprisingly long lives, too. Here are some randomly selected records, from shortest to longest known lifespan:

- Greater Roadrunner — 3 years
- Varied Thrush — 5 years
- Blackburnian Warbler — 8 years
- Wood Thrush — 8 years
- House Wren — 9 years
- Ruby-throated Hummingbird — 9 years
- American Goldfinch — 10 years
- Eastern Bluebird — 10 years
- Yellow Warbler — 10 years
- Baltimore Oriole — 11 years
- Chipping Sparrow — 11 years
- Downy Woodpecker — 11 years
- Black-capped Chickadee — 12 years
- Tree Swallow — 12 years
- American Robin — 13 years
- Great-crested Flycatcher — 13 years
- Tufted Titmouse — 13 years
- American Crow — 14 years
- American Kestrel — 14 years
- Northern Mockingbird — 14 years
- House Sparrow — 15 years

(continued)

(Long Live the Birds, *continued*)

▶	Northern Cardinal	15 years
▶	Brown-headed Cowbird	16 years
▶	Steller's Jay	16 years
▶	Turkey Vulture	16 years
▶	Blue Jay	17 years
▶	Gray Catbird	17 years
▶	Eastern Screech-Owl	20 years
▶	Great Blue Heron	24 years
▶	Mallard	26 years
▶	Great Horned Owl	27 years
▶	Ring-billed Gull	27 years
▶	Bald Eagle	30 years
▶	Canada Goose	30 years
▶	American White Pelican	31 years
▶	Mourning Dove	31 years
▶	Sandhill Crane	31 years

An American Robin can produce three successful broods in one year. On average though, only 40 percent of nests successfully produce young. Of those that fledge, only 25 percent survive to November. From that point on, about half of the robins alive in any year will make it to the next. Despite the fact that a lucky robin can live to be 13 years old, the entire population turns over on average every six years.

Q Why don't birds fall off branches as they sleep?

A When a bird bends its legs to perch, the stretched tendons of the lower leg automatically flex the foot around the branch and lock it in place. This involuntary action is called "the perching reflex."

^^

Coping with the Cold

Q How do birds stay warm all winter, especially when the temperature is below zero?

A Bird bodies are like a well-insulated cabin. Their outer, or *contour*, feathers keep moisture and wind out. Their inner, or *down*, feathers trap air, providing insulation to hold their body heat inside. Northern birds grow additional down feathers during the fall as temperatures start plunging. When they sleep, they raise these feathers, maximizing their insulation just as we do when we plump up a down sleeping bag or jacket.

Even the most well insulated cabin can be darned cold without a heat source within. Birds have two "furnaces" — their own muscle activity and their metabolism. Both burn calories

305

from their food to produce heat. Birds can survive northern winters only if they can get enough high-energy food to maintain their body temperature. This is especially tricky because the farther north they are, the longer the nights, so the less time each day can be spent searching for food to stay alive over the long night.

On cold nights, roosting chickadees "turn their thermostat down" by allowing their body temperature, normally above 100°F (38°C), to drop to as low as 82° (28°C). This saves energy, just as we do when we turn down the thermostat in our house or apartment. When a chickadee awakens on a cold winter morning, it immediately starts to shiver. That muscle activity quickly raises its body temperature back to normal.

Another strategy that some species use to survive long winter nights in the far north is to cozy up to one another. Chickadees don't do this — each one sleeps in its own cavity — but bluebirds, creepers, and Pygmy Nuthatches sometimes do.

Redpolls have more rods in their retinas than do many songbirds, allowing them to see better in dim light. They start feeding before dawn and continue to eat voraciously after sunset. They also have well-developed pouches along their esophagus, allowing them to load up on seeds at day's end so they can stoke the metabolic fires through the night.

Q Why don't birds get cold feet?

A Actually, songbirds do get very cold feet: the surface temperature of their toes may be barely above freezing even as the bird maintains its core body temperature above 100°F (38°C). But most birds don't succumb to frostbite because there is so little fluid in the cells of their feet, and because their circulation is so fast that blood doesn't remain in the feet long enough to freeze.

We don't know if cold feet bother birds. We do know that they have few pain receptors in their feet, and the circulation in their legs and feet is a double-shunt — the blood vessels going to and from the feet are very close together, so blood flowing back to the body is warmed by blood flowing to the feet. The newly cooled blood in the feet lowers heat loss from the feet, and the warmed blood flowing back into the body prevents the bird from becoming chilled.

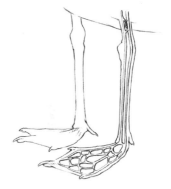

BABY, IT'S COLD

I've lived in northern Minnesota for over a quarter of a century, marveling at the winters. The coldest temperature ever measured in the state, –60°F (–51°C), was recorded in Tower, Minnesota, on February 2, 1996. Anticipating a record-breaker, one man slept out in a snow fort that night and emerged triumphant in the morning, clad in high-tech garb, to the cheers of hardy observers, TV cameras, and boom-microphones. No one seemed to notice the Black-capped Chickadees calling in the background. Each one of them had spent that record-breaking night sleeping outdoors, too, each alone in a small tree cavity, naked as a jaybird.

CHAPTER 10

Of Eagle Eyes and Ultrasound: How Birds Perceive the World

Vision that extends into the ultraviolet. Hearing that extends to frequencies well above what the keenest teenager can perceive. Most birds have senses of touch, taste, and smell that may be quite similar to our own, or poorer, but they can also perceive the earth's magnetism and polarized light. How well can they really see and hear? Those are just some of the questions people ask about how birds experience the world.

A Bird's Eye View

Q How good are "eagle eyes"?

A Birds in general have extraordinarily good eyesight. Imagine how acute a hawk's vision must be if it can soar hundreds of feet in the air, scanning the ground for the slightest movement to home in on a mouse.

Bird eyes are bell-shaped, with a large retina. A 3-pound Great Horned Owl's retina is larger than an adult human's. Bird eyes move very little, and so the only parts that are not covered by skin and feathers are the iris and pupil. But beneath the surface, their eyes are huge. Eagle eyes are as large as or larger than human eyes, and in some species the combined weight of the eyes is heavier than the brain!

Rather than being shaped like mammalian "eyeballs," avian eyes are concave in back, giving them a relatively larger retina than humans have. Bird eyes also have a distinctive feature: a large projection from the rear surface of the eye near the optic nerve, called the *pecten oculi*. Scientists don't yet understand why birds have this, but it probably nourishes the retina.

We know that birds' visual acuity is far better than ours, allowing them to perceive objects much smaller and farther away than we can. The cones — the cells in the retina responsible for visual acuity — are packed about five times more densely in areas of a raptor's retina than in ours. There are about a million cone cells per square millimeter!

In birds, each eye sends information to just one half of the brain, allowing birds to process information from separate visual fields independently, but an area of the brain called the "Wulst" processes the information from both eyes to provide stereoscopic vision, the means by which birds have depth perception, which is critical when an eagle or heron is homing in on a fish or when a flycatcher is snapping up a moth.

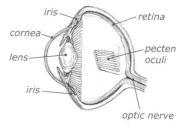

~ ~

Q Do birds see colors?

A Yes, they do, and they can even see some wavelengths in the light spectrum that are invisible to humans. Birds use their excellent color vision to find food, such as ripe fruits and flowers. Their colorful plumage is also important in courtship. Many studies have shown that when given a choice, female birds often prefer males with the most colorful feathers.

Birds can detect polarized light that humans can't see. Experiments show that pigeons and migratory songbirds use polarized light as a cue to help them navigate in the right direction.

Birds can also see ultraviolet light, which is invisible to us. Feathers reflect in the ultraviolet spectrum, and birds can see and use this information to help them discern sex and age differences in one another and even recognize individuals. In experiments in which female flycatchers could choose between a male with normal feathers and one treated with sunscreen to block the UV-reflection of his feathers, females showed a strong preference for normal, ultraviolet-reflecting feathers.

Many birds, including seabirds such as terns, gulls, and albatrosses, have red or yellow oil droplets, containing high concentrations of carotenoids (the same pigments that make carrots orange) within the cone cells in their eyes. Light travels through the oil droplets before it reaches the visual pigments. The oil droplets filter out some wavelengths of light, narrowing the color range that each cone perceives, which means the cones perceive colors more accurately than we can; birds with these oil droplets apparently also see better in hazy or watery conditions.

The ability to see ultraviolet light also comes in handy for finding food. Kestrels can actually see where their prey — small rodents called voles — have been walking. Fresh trails reflect UV light from urine that the voles have left behind, so kestrels can detect which trails are most likely to lead to a meal. How do we know this? Finnish researchers put out fresh trails of rodent urine in an experimental field and found that Eurasian Kestrels hunted along these trails. In the laboratory, kestrels spent more time hovering above and inspecting urine-treated paths illuminated by UV light than those illuminated by artificial white light.

✓ ✓

Q **I read that Peregrine Falcons can dive toward a duck at speeds of over 100 miles per hour. How do they protect their eyes from dust and insects, and from simply drying out, at such high speeds?**

A You're right that keeping eyes lubricated and dust free are important issues for birds. Birds have a translucent, semitransparent inner eyelid called the nictitating membrane that sweeps across the surface of the eye, allowing them to blink without blocking their view. Some birds, especially diving species such as loons and some ducks, have a clear "window" in the middle of the nictitating membrane that they probably use as we do goggles to improve underwater vision.

In addition, birds have two different tear glands. The lacrimal gland has many tear ducts along the lower eyelid. There is a second tear gland at the base of the nictitating membrane to maximize lubrication as that membrane blinks.

v v v v v v v v v v v v v v v v v v v v v v v v v

Q I spend a lot of time watching birds. On TV I see hawks and owls holding their heads upside down when looking at something, but I never see them do that in nature. Why?

A Virtually all birds and lizards, and also primates, have an area of the retina called the *fovea*, where vision is especially keen. The fovea of hawks and owls is above the midline, so from high perches or in flight they can see objects far below especially keenly. This is important because their food is normally found below, and they seldom need to see objects above their heads as clearly. Those TV hawks and owls holding their heads upside down are almost always captive birds, perched much lower than they would be in the wild in the presence of a photographer. When something looms above a bird's normal field of view and the bird can't change its position, it holds its head upside down to get the object in the fovea.

v v v v v v v v v v v v v v v v v v v v v v v v v

Q When robins cock their heads toward the ground, are they listening for worms?

A Nope, they're looking for them. A scientist named Frank Heppner designed an elegant series of experiments to establish that vibrations, odors, and sounds do not help robins find worms. Instead, robins see worms inside their subterranean burrows by peeking into the tiny holes at the surface, or they watch for them wiggling on the ground. Robins cock their heads to focus close up with one eye. The pupil of the eye looking up is adjusted for daylight while the one focused down is open wider for seeing inside the dark little holes.

Male robins sing most intensely while it's still too dark to see worms, and females lay their eggs at midmorning rather than at dawn when most songbirds do. This enables them to focus their attention entirely on feeding during the time of day when worms are most visible, at first light before the sun sends them underground for the day.

The other bird most specialized for feeding on earthworms is the American Woodcock, a plump shorebird with a very long bill. Like robins, woodcocks pick up any earthworms they might see wriggling about on the surface. Woodcocks probably detect underground worms at least partly by touch, probing beneath the soil with the sensitive tip of their long bill.

ˇ ˇ

Hear and There

Q **Why are owls the only birds with visible ears?**

A All birds have ears, usually hidden behind feathers on the sides of their face. The "ear tufts" on owls aren't ears at all, but feathers sticking up on their heads. Ornithologists speculate that when the feather tufts are raised, typically when an owl is alarmed, they make the bird look somewhat like a broken branch, which may help the owl avoid detection.

The feather tufts may also help owls recognize and visually communicate with one another. A few people think the feathers may enhance their sense of touch around their face, since every feather is attached to a nerve. These ear tufts also enhance many yellow-eyed owls' catlike appearance, and so may give them a second or two of extra time to escape when approached by a predator, because even large predators are at least a little intimidated by the claws, teeth, and fierce fighting ability of cats.

ˇ ˇ

Q **How well do birds hear?**

A Birds have more acute hearing than we do. Although audiology tests indicate that we can hear higher frequencies than the birds tested in some experiments, the range at which we hear best is lower than that of some songbirds, and they can resolve notes in songs that are too rapid for us to easily distinguish. Young humans can usually hear from roughly 20 to 20,000 hertz (Hz). (One hertz equals one vibration per second. To give you a sense of what that means, the lowest key on an 88-note piano keyboard produces a sound at 27.5 hertz, Middle C is at 261.6 hertz, and the highest note is at 4,186 hertz.) As another point of comparison, dogs can hear sounds in the 67 to 45,000 Hz range, while cats have the keenest ears, capable of distinguishing sounds from 45 to 64,000 Hz.

Two species of grouse from Europe and Asia, called capercailles, can produce sounds below 20 Hz in their breeding displays, and North America's Ruffed Grouse drums at 40 Hz. They produce these sounds to attract mates and to defend territories from rivals. Behavioral studies show that pigeons can detect sounds as low as 0.05 Hz. This may provide one of the cues that pigeons use to home, because such sounds as wind blowing over different kinds of terrain and ocean waves can carry over hundreds or even thousands of miles. Great Horned Owls in some lab experiments appear to hear better than we do at the lowest frequencies, but they may be perceiving these sounds through their sense of touch as we perceive the lowest bass sounds by feeling the vibrations. Grouse may have evolved to produce sounds at the limits of owl hearing abilities so that they can safely display at dawn, dusk, and night, when owls are most actively hunting, preferably for something as large and tasty as a grouse.

Most songbirds seem to hear best in the range between 1,000 and 5,000 Hz, about the range of the top two octaves on a piano and a bit higher. Most people can easily hear songbirds singing at the mid-range, but as we grow older, we start losing the highest and lowest frequencies. Some bird songs are simply too high for most older people to detect without hearing aids: the sibilant trills of Cedar Waxwings in the 6,000 to 9,000 Hz range, the sweet notes of the Cape May Warbler at 10,000 Hz, and the lovely but sky-high Blackburnian Warbler songs that ascend at the end to sometimes top 11,000 Hz.

ˇ ˇ

Q Do birds use sonar?

A Sonar, or echolocation, in the animal world is a method for detecting and locating objects by sending out sound waves that are reflected back by the objects; it's usually found in nocturnal creatures or those that hunt in the blackness of the ocean deep or in dark caves. Sperm whales use sonar to find and catch squid in the ocean depths, and bats use it to catch insects and to avoid colliding with objects. Bats emit extremely

> Birds can distinguish the individual notes of complex, rapid songs much more easily than we can — to say nothing of being able to produce those songs in the first place. A Winter Wren in western North America sings an average of 36 notes per second! We can only resolve those individual notes by playing back a recording of the song at slow speed.

high frequency sounds; these ultrasound wavelengths are tiny enough to bounce off very small insects.

Among birds, the Oilbird — a nocturnal species of Trinidad and northern South America — and cave swiftlets of Southeast Asia are known to use echolocation. These birds produce audible clicks that bounce off the cave walls and other impediments, such as stalactites and stalagmites, allowing the birds to safely negotiate their treacherous homes. But neither the swiftlets nor the Oilbird can produce ultrasound notes, so their sonar is at best only one-tenth as functional as that of bats — Oilbirds can detect items only larger than 20 millimeters in diameter and will collide with anything smaller.

^v^v^v^v^v^v^v^v^v^v^v^v^v^v^v^v^v^v^v^v^v^v^v^v^v^v^v^v^v^v^v^v^v

The Nostrils Know

Q **Can birds smell?**

A People once believed that, with very few exceptions, birds couldn't smell at all. They don't have a specialized nose but simply nostrils, necessary for breathing and usually located near the base of their upper beak. But some species, including several ground birds and also some North American vultures and marine species called "tubenoses," do have fairly large olfaction (smell) centers in their brains.

Many "tubenose" seabirds, especially petrels, shearwaters, and fulmars, locate their oceanic food by smell. Petrels are attracted specifically to the smell of dimethyl sulfide, an aromatic substance released by microscopic algae when tiny drifting invertebrate animals called zooplankton are feeding on them. The petrels don't eat the algae; they're after the zooplankton. This zooplankton isn't easily seen, and petrels feed by night as well as by day, so their sense of smell is very useful, and very keen.

Tiny Wilson's Storm-Petrels, smaller than American Robins, can detect slicks of dimethyl sulfide from great distances. Their albatross relatives also have a good sense of smell but are not attracted to dimethyl sulfide, probably because these larger birds feed on fish and squids rather than plankton.

Recent studies have shown that even some songbirds, with relatively tiny olfaction centers in their brains, can smell. For example, Cedar Waxwings, which eat berries that can ferment and make them sick, have a better sense of smell than Tree Swallows, which probably can't take in the odors of flying insects as they snap and swallow them in flight.

Kiwis (flightless birds from New Zealand) are the only birds with nostrils close to the tip of their bills. Their olfactory centers are about 10 times the size of those of other birds relative to their size, and they apparently find the earthworms they feed on by smell.

Some homing pigeons use their sense of smell as one cue for navigating home. And some seabirds use their sense of smell to locate their nest.

Great Horned Owls eat skunks. Their sense of smell hasn't been evaluated carefully, but if they are good at smelling, apparently their sensibilities about what smells good are quite different from ours!

Q Do vultures find dead animals by smell or by tracking predators or other scavengers on the ground?

A Researchers proved fairly long ago that Turkey Vultures can smell. In 1938, the Union Oil Company discovered that by injecting a strong-smelling organic chemical called mercaptan into gas lines, they could readily find leaks by monitoring vulture activity above the pipelines. Some mercaptans smell like rotting cabbage or eggs. They and related chemicals are released as carcasses decompose. To us, mercaptans smell horrible, but for vultures they are associated with fine dining.

In a 1986 study in Panama, Turkey Vultures found 71 of 74 chicken carcasses within three days. There was no time difference between finding concealed and unconcealed carcasses, and the only carcasses the vultures seemingly had trouble finding were the freshest ones. Even though the older carcasses emitted a stronger odor, the vultures showed a definite preference for eating fresher carcasses.

Greater and Lesser Yellow-headed Vultures of Central and South America, which are closely related to Turkey Vultures, seem to have comparable reliance on their sense of smell for finding food, and King Vultures may also use smell to find food. These species must all be able to find carrion in forests where the canopy visually obscures dead animals. Unlike these species, Black Vultures, which find their food primarily in open country, depend far more on vision and are believed to have a relatively poor sense of smell. Of course, one strategy that all vultures use to locate food is to watch for other circling vultures to drop down suddenly; in that sense, even Turkey Vultures find much of their food visually.

BAD-TASTING MEDICINE

Some strong tastes elicit interesting responses in birds. I once watched a fledgling Blue Jay that I was rehabbing pick up a large ant with the tip of its beak and then open wide to pull it into its mouth. Suddenly its crest went up and it spit out the insect and shook its head hard, all the time running its lance-shaped tongue against the roof of its mouth. Suddenly it picked up the ant again and started smearing the insect against its feathers.

Many birds engage in "anting." Ant bodies are covered with a bitter chemical called formic acid, which may afford birds some protection from mites and lice. One European researcher studying a population of pipits that were suffering from a heavy infestation of feather mites learned that mites on birds that had been anting suffered much higher mortality than the mites on non-anting birds. In the case of my Blue Jay, apparently the taste of the ant elicited the behavior. People have reported birds anting with other items, including mothballs, cigarette butts, and onions.

Q How well do birds taste their food?

A Bird taste buds are similar in structure to mammalian ones, but they have significantly fewer than we have. Chickens have 24 taste buds and pigeons have fewer than 60, whereas we humans have about 10,000 and rabbits have about 17,000! Most of our taste buds are on our tongues, but birds have very few on the tongue, and none at all on the tip. Instead, most of their taste buds are on the roof of the mouth and deep in the oral cavity.

One researcher fed bread mixed with quinine to parrots, and they ate it without any apparent objections. In taste preference tests on pigeons, one researcher found that the birds rejected sour or bitter solutions, preferred low concentrations of salt and high concentrations of sucrose, but didn't seem to respond to glucose at all. In another experiment, pigeons didn't seem to have any response to quinine, and half responded to saccharin.

Birds do respond to some bitter tastes. If a Blue Jay bites into a monarch butterfly, the strong bitter taste makes the jay spit it out. Monarch caterpillars feed on milkweed, and foul-tasting toxins from the milkweed, called *cardenolide aglycones,* are taken up by the caterpillar's tissues, remaining in the adult butterfly. The bright orange color of monarchs protects them from any bird that tasted one and either found it too bad-tasting to eat or swallowed it and got sick.

Extra-Sensory Perceptions

Q Do birds have a sense of touch?

A Many songbirds have very few tactile nerve endings on their feet. Gray Jays have been known to stand on cast iron pans where bacon is sizzling without apparent discomfort! But birds have an excellent sense of touch that works in other ways. Beneath the feathers, bird skin is very sensitive, especially where the flight feathers attach and at the wing joints, allowing birds to sense and adjust to the tiniest changes as they fly. Thanks to their exquisite sense of touch, hawks and other soaring birds can feel the rising air of a thermal air current to capitalize on easy lift during migration.

Furthermore, bird beaks and tongues may not have many taste buds but they do have a great many touch receptors, as shown by the following examples.

▶ Woodcocks, snipes, and some sandpipers have exquisitely sensitive tactile sensory receptors at the tips of their bills, providing them with the means to detect and grasp worms more than two inches below the surface by touch alone.

▶ Woodpeckers have dense, sensitive touch receptors at the tips of their tongues, which they use to feel out insects dwelling deep in wood.

▶ Finches have a great many touch receptors in their beaks and tongues, allowing them to hold and crack open a seed shell in the corner of the beak as they deftly manipulate and swallow the seed inside.

▶ The bills of ducks and geese are extremely sensitive, especially at the tip and along the outer edges. Just along the edge of the palate, a Mallard has 27 touch corpuscles per square millimeter, which compares with 23 touch corpuscles per square millimeter in the most sensitive part of a person's index finger.

▶ The bills of Wood Storks are so sensitive that when blindfolded, these birds could close their beaks on a live fish in as little as 0.019 seconds after first touching it; this is less than half the time it takes for us to blink.

▶ The feathers around the mouths of Whip-poor-wills, called *semibristles*, have a tactile function similar to that of a cat's whiskers and may also increase the useful area of the gape for sweeping up insects.

∨ ∨

Q **Can birds really feel barometric pressure?**

A We have an abundance of field data that implies that birds must be able to feel barometric pressure. Before storms, as pressure is falling, birds feed more actively than at other times. During nocturnal migration or in foggy conditions, birds can maintain a safe altitude for long distances without being able to see or visually gauge the distance to the ground beneath them. Both of these abilities seem to depend on being able to detect barometric pressure, but scientists haven't yet been able to discover exactly how birds do this. However, scientists have proven experimentally that homing pigeons are extremely sensitive to small changes in air pressure, comparable to about a 25-foot difference in altitude.

⌄ ⌄

Q **Do birds have any senses that we humans lack entirely?**

A Many birds can sense the earth's magnetic pull and use this information to help them navigate. In one experiment, scientists fitted homing pigeons with tiny metal helmets. Half of the birds wore magnetic helmets of iron and the other half wore helmets made of a nonmagnetic substance of the same weight. The birds were taken away from their home loft and released.

The pigeons with magnetic helmets usually became lost on cloudy days, probably because the helmet interfered with their ability to perceive the earth's magnetic field, which they use to help them navigate when they can't see the sun. (When the sun came out, pigeons with magnetic helmets found their way home, too.)

Tiny crystals of magnetite have been found near the olfactory nerves of pigeons between their eyes, and magnetite has also been found in similar tissues or the upper beaks in many migratory species. In a recent experiment, scientists trained caged pigeons to hop to one end of their cage when the magnetic field was normal and to the other end when the scientists switched to an abnormal magnetic field. The birds could only learn this task if, indeed, they could detect differences in the magnetic field.

When scientists attached magnets to the pigeons' upper beaks, the birds could no longer perform the task. And when scientists temporarily froze their olfactory cavity (where deposits of magnetite are concentrated), the birds also failed to demonstrate this behavior.

Some birds may be able to detect magnetism by way of light receptors in their eyes that may be able to convert light and magnetic fields to nerve impulses. However they do it, the world of a bird must be filled with sensations that we can only imagine!

During the breeding season, you might see three Mourning Doves flying in tight formation, one after another. This is a form of social display. Typically the bird in the lead is the male of a mated pair. The second bird is an unmated male chasing his rival from the area where he hopes to nest. The third is the female of the mated pair, which seems to go along for the ride.

CHAPTER 11

On Angels' Wings:
Fascinating Feathers

We mammals are a varied class of animals coming in myriad shapes and sizes, from bats to whales, giraffes to platypuses, lions to human beings. The tiniest shrews are of similar size to hummingbirds, but the largest whales are orders of magnitude larger than the biggest ostriches or even the most gigantic birds that ever lived, the moas and elephant birds. All birds stand or perch on two feet, and although a few, like the kiwi and ostrich, have only vestigial wings, the overall bird shape is distinctive.

Bird bodies are covered by and get their distinctive shape from feathers — a unique feature found nowhere else in the animal kingdom. What mammal could truly be called "resplendent" or "scintillant," or be justifiably named for any of the gems common in hummingbird names — ruby, amethyst, topaz, emerald, sapphire, or garnet?

Small wonder our depictions of angels portray them bearing the wings of birds. And small wonder that when we see birds, our minds are filled with questions.

Fine Feathered Facts

Q **What are feathers made of?**

A Feathers are features unique to and universally found in all species of birds. Each one grows out of a specialized structure called a *papilla*. Human hair also grows out of papillae — biologists are still researching how similar these structures may be, and how similar feathers may be to reptilian scales. Compared to scales, feathers are much softer and more flexible, are extremely lightweight but strong, and provide excellent insulation against excessive cold and heat.

Intuitively, it seems like feathers must have started out as scales that evolved to become frayed. But developmental biologists point out that feathers begin growing as tubular structures quite different from scales. Scientists are still debating *why* feathers evolved — whether they were originally for insulation, flight, display, or some other function. There are lots of fossils that suggest many feathered dinosaurs did not fly.

Feathers are composed of phi-keratin, which is not known to occur in any nonbird animal tissue except alligator claws.

The largest feathers are peacock tail feathers, which can be as long as 5 feet (1.5 m). The tiniest feathers are the eyelid feathers of the Bee Hummingbird of Cuba, measuring a mere 1/63 of an inch (0.40 mm).

Q How many feathers does the average bird have?

A People have not counted feathers on a great many species, but during the 1930s, a few scientists took the time to count the contour feathers of several dead birds. (Contour feathers include the external body feathers but not other feathers, such as down.) They had to delicately pluck the feathers and count them one by one.

One Ruby-throated Hummingbird had 940 contour feathers, and one Tundra Swan had 25,216. The latter, so far, represents the most feathers ever counted on an individual bird. Fully 80 percent of them (20,177 to be exact) were on the head and neck — the parts of a swan that are submerged in water when the bird is feeding. Those are very tiny compared with the contour feathers on the rest of the body. Imagine counting them all!

The number of feathers may also vary seasonally. One researcher counted the feathers on a few White-throated Sparrows. Those collected in winter averaged 2,500 contour feathers; those collected in summer averaged only 1,500.

FEATHERS: NOT JUST FOR FLYING

A remarkable diversity of feathers exists for many purposes.

Flight: The feathers of the wings are strong, stiff, and lightweight enough to propel and provide lift to the bird in flight, and feathers of the tail are strong and stiff enough to help even the fastest falcons steer. The feathers on the head and body are arranged to streamline the body for flight and when facing a stiff wind.

Protection: Outer feathers give birds a fairly waterproof shell that protects their skin from rain and snow, as well as sunburn. Beneath the outer feathers, down feathers trap and hold in body heat. Some body downs are adapted for insulating not just the bird but also her eggs, which do not generate their own heat. Many female waterfowl pluck down feathers from their bellies to line their nests.

In the most famous case, the exquisite down of female eiders insulates the eggs from the permafrost below even as the incubating duck protects them from frigid air temperatures above. When she leaves the nest to feed, she blankets the eggs in this down to hide them and keep them warm and dry.

Feathers often help camouflage birds, such as the mottled brown plumage of an American Woodcock that blends in with the forest floor.

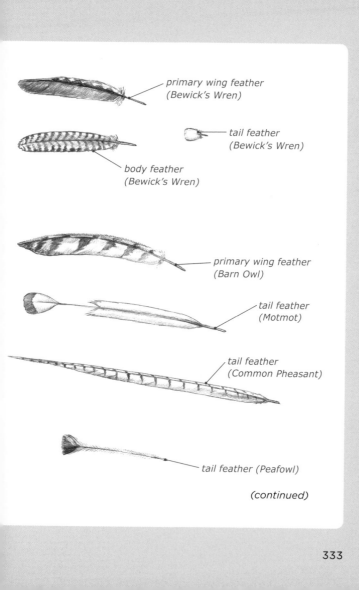

primary wing feather
(Bewick's Wren)

tail feather
(Bewick's Wren)

body feather
(Bewick's Wren)

primary wing feather
(Barn Owl)

tail feather
(Motmot)

tail feather
(Common Pheasant)

tail feather (Peafowl)

(continued)

(Feathers: Not Just for Flying, *continued*)

Attracting a Mate: Color patterns on feathers can have the opposite function of camouflage. Some brightly colored plumage makes a bird highly visible. Many birds have specialized feathers that appeal to mates.

The back feathers of many species of egrets and herons in breeding plumage are long and wispy. The barbules of the long tail feathers of male Indian Peafowl are decorated with a beautiful "eye." The Resplendent Quetzal has tail streamers (actually elongated rump feathers) that may be nearly twice as long as its body.

Special Functions: The tail feathers that brace woodpeckers, swifts, and some other birds against trees or other hard structures are very stiff.

Some insect-eating birds have specialized feathers called *rictal bristles* near their beaks, with an extremely stiff shaft and no barbs. These may help funnel insects into the bird's mouth or provide extra tactile sensitivity as do cat whiskers. Experimental data indicates that they also protect the bird's eyes from debris as it captures moths and other large, scaly insects in flight. Many birds have modified wing or tail feathers that allow them to produce sounds for communication during courtship displays.

Let's See Some Identification

Q I was walking with one of my friends when we came upon a random yellow feather. He looked at it for just a second and said it was a flight feather from the left wing of a Northern Flicker. How could he possibly know that?

A Your friend lucked into finding one of the easiest kinds of feathers to identify. He could tell it was a flicker's feather by the brilliant yellow color on the underside. No other species except the Gilded Flicker, found in Southwestern deserts, shares this. (Feathers from Northern Flickers from the West are red rather than yellow.) He could tell it was a flight feather by its stiff shaft and its length: flight feathers, whether from the wing or the tail, are usually longer than they are wide.

How did he know it was a feather from the wing rather than from the tail? All woodpecker tail feathers are straight, fairly symmetric, exceptionally stiff, and the tip tapers to a point. Wing feathers are more typically feather-shaped, have a very slight curve, and the two *vanes* (the webbed sides of the feather) are asymmetric: the leading edge of a primary wing feather is narrower than the trailing edge.

So your friend looked at the color and shape to deduce it was a flicker's wing feather, and put the yellow side down to see which way the feather curved and which was the narrow leading edge to tell whether it came from the right or left wing.

˅ ˅

INSPECTOR CLAW-SEU

🐦 **As with any fibers found at a crime scene,** feathers can provide important circumstantial evidence in murder investigations.

Prominent Federal Bureau of Investigation (FBI) hair and fiber expert Douglas Deedrick worked extensively with the world's most famous feather authority, the late Roxie Laybourne, to develop a permanent slide mounting medium that made it possible to establish microscopic data on tens of thousands of feathers. She in turn helped him for decades by examining and identifying feather evidence for FBI cases.

Although feathers aren't found in nearly as many investigations as other fibers (including hair), her services were used in as many as a dozen robbery, kidnapping, and murder cases each year. She also identified feathers implicated in airplane collisions with birds. Mrs. Laybourne spent her career with the Smithsonian Museum of Natural History, where she could compare feather samples with the huge collection housed there to verify her identifications.

Q **Is there a book that teaches how to identify feathers?**

A No. Some birds, such as flickers and Blue Jays, have many easily recognized feathers, and some birds have at least a few easy ones. For example, Cedar Waxwing tail feathers are each tipped at the terminal end with a brilliant band of yellow, and Eastern Kingbird tail feathers are tipped with white. But identifying most individual feathers is not nearly that straight-forward.

Most owl flight feathers are extremely soft, as if coated with velvet, and their leading primary wing feathers have a stiff "comb" on the leading edge. But it's tricky to know which owl a given feather might have come from, though size gives some hints. Ducks, shorebirds, and some other water birds have outer feathers that feel waxy. The better you get at bird identification, the more clues you'll be able to integrate to puzzle out individual feathers. But some feathers may be only identifiable by experts in laboratories.

~~~~~~

Colorful Characters

Q **Someone told me there's no such thing as blue feathers. Is this true?**

A Anyone who has ever observed a bluebird or jay knows darned well there are blue feathers. But there is no such thing as blue *pigment* in feathers. Blue in feathers is a "structural color," which means the color we see is due to a very special arrangement of keratin and air overlying a blackish pigment. When light bounces off this layer, it appears blue.

If you find a blue feather, check it out! When you hold it so the light bounces off it, it will be bright blue. But when you hold it so it's backlit, either by the sun or any artificial light, some of the pigment color — a dark brownish gray — will show through.

Budgerigars living in the wilds of interior Australia are mostly green, which is the result of yellow pigments and blue structure. Birds in captivity have been bred for different colors. If a bird's feathers lack the yellow pigment but maintain the blue structure, they'll be blue. If they lack the structure but maintain the pigment, they'll be yellow. And if they're lacking both the structure and the pigment, they'll be white.

Iridescence is also caused by feather structure. If you were to grind up a hummingbird's brilliant throat feathers or the metallic green feathers of a Mallard, you'd end up with a dark gray powder, the color of the feathers' pigment. But if you ground up some red feathers from a Scarlet Tanager, the powder would be red because those feathers are red from pigment.

˅ ˅

Q Early last spring, I saw a bird that was shaped just like a robin running on my front lawn. But it was all white! What was it?

A It was a robin, but unlike most robins, it was white because it was an albino, lacking the pigment *melanin.*

Birds that are normally or seasonally all white, such as swans, egrets, and ptarmigans, are not considered albinos, except in the extremely rare situation in which one also lacks pigment on the beak, feet, and eyes. If your robin had pink eyes, and a pale pinkish bill it was a complete albino. Unfortunately, because complete albinos lack pigment in their eyes, they have no protection from ultraviolet light and tend to become blind, dramatically shortening their lives.

Albinism is rare: one researcher estimated that only about one in 1,800 birds is an albino. In humans, the incidence of albinism is about one in 17,000. Albinism usually results from a genetic mutation that interferes with the production of tyrosinase, the enzyme necessary to produce melanin. Some birds are partially albino, with only certain feathers lacking melanin.

Few if any studies of the distribution or frequency of albinism in birds have been conducted. Some researchers have noted that American Robins and House Sparrows have the highest incidence of albinism of any birds, but this may well be because these two species are both conspicuous in backyards and other habitats where people spend a lot of time, making albinos more likely to be noticed and reported.

˅ ˅

Q I love the little red finches that nest in my begonia basket. For some reason, last spring the male was orange, not red, but their babies seemed healthy. What caused his unusual color?

A It was probably caused by his diet. House Finch colors are produced by carotenoid pigments that are incorporated into growing feathers if the birds are eating foods rich in carotenoids, which are found in a variety of wild fruits. In laboratories, when House Finches are fed a simple seed diet, the feathers that grow in after they molt are fairly dull and yellowish. When beta-carotene is added to the diet, the feathers that grow in after they molt are pale orange. And when a red carotenoid found in some fruits called *canthaxanthin* is added, their next feathers grow in bright red.

In some areas where House Finches have been introduced, such as Hawaii, their natural diet is poor in these pigments and the birds have dull plumage. In other areas, such as Michigan, their feathers can be brilliant.

Brighter males are more attractive to females. The dietary carotenoids that give males these bright colors also give females a good idea of which males are able to find the highest quality food, indicating directly which will be the best providers for their young.

v v v v v v v v v v v v v v v v v v v v v v v v v v v

Q I read that Sandhill Cranes don't have feathers on their faces, but when I saw one up close, those sure looked like red feathers! What gives?

A What you read was true, although the face does indeed look like it's covered with red feathers, because the skin of the forehead and crown of both sexes is covered with bumpy structures called *papilla* that look very much like tiny feathers. Blood vessels run very close to the skin's surface, and when the cranes are engaging in sexual or territorial displays, the papilla become engorged with blood, appearing brilliant red. To further confuse the matter, fine, very thin black plumes lie on top of the red skin, contributing to the feathery appearance.

Brightly colored skin on faces and heads of birds is not uncommon.

^v^v^v^v^v^v^v^v^v^v^v^v^v^v^v^v^v^v^v^v^v^v^v^v^v^v^v^v^v^v^v

Health and Beauty Aids

Q I just saw a totally bald-headed Blue Jay! At least I think it was a Blue Jay. Why was it missing all its head feathers?

A When some individual birds molt, their head feathers drop out nearly simultaneously. No one has figured out why this happens so frequently in Blue Jays and Northern Cardinals, especially because some individual cardinals and jays seem to molt only a few head feathers at a time, as do other species. Is it significant that these two unrelated species are crested? This is very unlikely, because some crestless birds, such as grackles, sometimes lose all their head feathers, too.

Some people have speculated that bald birds are plagued with mites, but this is unlikely. For several years I had two Blue Jays housed side by side in my rehab facility. Every fall one of them, Sneakers, lost all its head feathers at once while its neighbor, BJ, with whom it shared food, never molted all its head feathers. If one of them had mites, the other probably would have. It seemed to be an individual quirk.

˅ ˅

Q **Birds always look worse when they leave my birdbath than when they arrived! Why do they bother?**

A Wet feathers do indeed look scraggly! Birds bathe to get clean, but when they step out of the bath, their feathers look as unkempt as our hair does just after washing. Birds dip in the water while opening their feathers. If they're in a completely secluded area and haven't detected any potential predators lurking about, they may splash and shake and soak through the outer feathers to the down and skin beneath. Then they fly off to preen in a safe hiding spot, nibbling off remaining dirt, parasites, and oils. It is astonishing how quickly the feathers dry and how automatically they resume their shape.

˅ ˅

FEATHER CARE

Feathers are a wonderful adaptation for flying creatures. If a bat flying through trees accidentally tore a wing on a branch, the injury could permanently ground it. If a bird were in the same kind of collision, the branch might damage a feather or two but more likely would simply part two feathers, which the bird could realign simply by preening.

Basic Grooming

As wonderful as feathers are, they must be properly groomed to hold up under the rigors of a bird's daily life. Birds nibble them to realign hooks and barbules to maintain the integrity of the vanes, and in the course of this preening they pick off lice and mites, too. To keep feathers supple and protected from sun, wind, rain, and saltwater damage, most birds have a gland at the base of the tail called a uropygial gland or "preen gland." During preening, birds nibble at this small, pimplelike structure to squeeze out drops of oil that they work into their feathers to keep them supple and in good condition, as hand lotion can do for our skin. The oil may also foster helpful species of fungus that protect

(Feather Care, *continued*)

feathers from parasites while helping control harmful fungus growth.

People once believed that the oil secreted by this preen gland directly waterproofed feathers, but when researchers removed the preen gland from ducks, they found that the feathers continued to be waterproof until, over time, they grew brittle and started breaking. Apparently the oil indirectly contributes to waterproofing not by repelling water itself but by helping the feathers to maintain their structure.

preen gland

Parasite Prevention

Mites and some other ectoparasites eat feathers, damaging hooks and barbules. Preening helps get rid of these pests but often isn't enough, so many birds engage in additional activities. In one, called "anting," birds apparently take advantage of the formic acid or another strong chemical in ants to repel mites. There are two techniques for anting. In passive anting, the bird sits still, almost as if in a trance, on an anthill or other spot where ants are swarming, allowing them to crawl through its feathers.

In active anting, the bird picks up an ant or other thing with strong chemicals, such as a mothball or cigarette butt, and smears it on its feathers. Sometimes after this active anting, the bird eats the remaining insect.

Birds often sunbathe. As while passively anting, sunning birds seem to enter into a trancelike state; they assume strange positions, often leaning to one side and opening spaces among all their feathers. By raising the temperature of their skin and feathers, they may be driving away parasites or somehow helping to maintain the condition of their feathers. Sun and wear usually take the heaviest toll on wing and tail feathers.

Birds also bathe in water, snow, or dust. It seems odd to us to imagine bathing in dirt, but snow and dust baths can scrape off a lot of parasites.

Drying Out

Anhingas, cormorants, pelicans, storks, vultures, and some other birds sometimes sit for long periods with their wings outstretched to dry them. Anhingas and cormorants spend a lot of time underwater chasing fish, and in both cases feathers can become quite wet, so this activity helps dry them. Anhinga feathers are very permeable to water, much more so than are cormorant feathers. This may help

(Feather Care, *continued*)

anhingas to swim low and sink into the water more easily, but because the feathers can become very waterlogged, the birds must hold their wings out for longer periods than do cormorants. Vultures often stretch their wings out in the morning. This may help dry feathers that had become wet from rain or dew overnight.

In all these ways, birds keep their feathers in condition, but even with careful maintenance feathers eventually degrade. So birds molt them, replacing each feather usually at least once, often twice, and in some cases three times a year.

Q A junco with a broken tail has been visiting my feeder. One of my friends said it would grow back, but the bird has been here for two months and it looks exactly the same as when it arrived. What is keeping it from growing a new tail?

A If your junco's tail feathers had been torn out, the bird would have replaced them immediately. However, broken and worn feathers that are intact where they meet the body aren't replaced automatically until the time of year when the bird would normally replace its feathers. The timing of molting varies among species; your junco should grow in new tail feathers in late summer.

Courting Red-tailed Hawks put on a display in which they soar in wide circles at a great height. The male dives steeply, then shoots up again at an angle nearly as steep. After several of these swoops, he approaches the female from above, extends his legs, and touches her briefly. Sometimes, the pair grab onto one other, clasp talons, and plummet in spirals toward the ground before pulling away from each other.

LIKE A ROCK

When I was rehabbing birds, one of my charges was
a baby Blue Jay. When he was newly fledged, he spent
most of his days outside in my yard. One afternoon when
a severe thunderstorm was looming, he was nowhere to be
found. When the storm hit, I had to go inside without him.
The moment the rain subsided, I headed out and searched
my neighborhood, calling for him. When he heard me, he
started squawking *MAAAA! MAAAAA!* from a power line just
above a busy bus stop about a block away.

When I reached him, he was sopping wet and bedrag-
gled, his crest and feathers plastered against his skin, looking nothing like a Blue Jay. The dozen people below waiting
for a bus stared and laughed as I called, "Come on down,
Ludwig!" He flapped his wings and let go of the wire, but his
sopping wet wings and tail provided absolutely no lift and
he dropped like a rock to the sidewalk, then hopped up to
me, still squawking *MAAAA! MAAAAA!*

The next time he got caught in a storm, he'd figured out
that hiding in a sheltered branch is far wiser than sitting out
in the open.

CHAPTER 12

Those Magnificent Flying Machines: How Birds Fly

Who among us hasn't wondered at, and perhaps even envied, the shape-shifting robin that one moment is hopping, two-legged, on the lawn and then suddenly jumps up, opens its wings, pulls in those legs, and flies! In eighth-century Spain, Abbas Ibn Firnas studied the dynamics of bird flight and carried out his own flying experiments. In sixteenth-century Italy, Leonardo da Vinci invented several flying machines, again closely studying the movements of birds.

Now we can take a jet to just about anywhere in the world, a feat impossible but for the inspiration and understanding of physics that came from looking at birds. But well over a millennium after the first human efforts at flight, we still watch flying birds with wonder.

^v^v^v^v^v^v^v^v^v^v^v^v^v^v^v^v^v^v^v^v^v^v^v^v^v^v^v^v^v^v^v^v

How Do They *Do* That?

Q How can birds stay up without flapping their wings?

A Depending on the shape of the wings, some birds must flap to stay aloft whereas some can glide or soar without flapping at all for many minutes or even hours.

Bird wings are shaped to form an *airfoil*. When a bird moves forward through the air, the shape and curve of the wing cause the air to flow faster above the wing than below it. The faster air above lowers the pressure (drawing the bird upward) while the

low wing loading, low aspect ratio (hawk)

medium wing loading, medium aspect ratio (shorebird)

medium wing loading, medium aspect ratio (grouse)

low wing loading, high aspect ratio (albatross)

slower air below raises the pressure (pushing the bird upward). This force holding the bird up is called *lift*, and it requires that the bird be moving forward or facing into a fairly stiff wind.

In order for a bird to soar rather than to flap, its weight must be low relative to the surface area of the wings and tail, giving it a low *wing loading.* Birds with low wing loading include cranes, hawks, vultures, anhingas, pelicans, and many others. Loons are an example of birds with high wing loading — that is, their body weight is very high relative to the surface area of their wings. If a loon is missing just a couple of flight feathers, flight becomes difficult or impossible. Penguins have such extremely high wing loading that their small wings can't get them off the ground at all. Hummingbirds and many songbirds have high wing loading and must keep flapping to stay aloft.

The shape of the wings determines how a bird soars. An albatross, with its extremely long, slender wings, has a high *aspect ratio,* which allows it to fly at high speeds at low altitudes (sometimes barely above the ocean's waves), while being extremely maneuverable. A condor's wide, slotted wings give it a low aspect ratio. Birds with a low aspect ratio are adapted to slow speed, high altitude, effortless flight. Birds such as these, with aspect ratios at either extreme, may have difficulty taking off from the ground without facing a stiff headwind. Most soaring birds fall somewhere between these extremes.

ˇ ˇ

Q **How can owls fly so silently?**

A Given their body size, owls have very large wings, allowing them to flap more slowly than many birds, reducing noise. The velvety surface of their flight feathers, along with their fluffy body plumage, helps absorb sound waves. The velvety surface also muffles any scraping sounds as the feathers slide against each other in flight.

But the one feature that is probably most significant in creating an owl's silent flight is the stiff, comblike fringe along the edge of the outer vanes of certain wing feathers, the first primary feathers. This softens the contact between the air and the leading edge of the wing, essentially breaking up the *whoosh* of a wing beat into 50 or 100 tiny *whoosh*es.

MADE FOR FLIGHT

Birds are masters of flight, and because we can easily see their wings and feathers, many people assume that those are the key features that enable birds to fly. But birds are actually built for flight on the inside, too, including their skeletons. The bones of birds are exceedingly light: most flying birds have skeletons that weigh just half or a third of the weight of their feathers. Flying birds have huge airspaces in many or most of their bones, although some bones in each bird are filled with enough bone marrow to manufacture blood cells. The long wing bones of vultures and swans are hollow enough that in earlier times they were made into primitive flutes, which have since been recovered from caves. Charles Darwin had a pipe whose stem was made from the hollow wing of an albatross.

Many of the avian bones that are equivalent in origin to human bones are reduced and fused into forms unlike our own. For example, the bones corresponding to our finger bones have been fused and modified to support all the primary flight feathers. Birds also have some vertebrae that are fused together to form a rigid plate of bone that widens laterally to fuse with the pelvic bones. The entire structure makes a light, stiff framework that allows a bird's legs to support its body with a minimum of heavy muscles.

Flying birds have a huge, keel-shaped sternum protecting the chest and part of the belly from physical blows while providing a perfectly placed surface for attaching the large wing muscles. The backbone, ribs, and sternum together form a flexible but strong box, housing and protecting the heart, lungs, and other major organs.

353

Some birds such as woodpeckers and blackbirds may cruise forward on a powerful flap or two and then soar for a moment, wings folded, before the next flap. This gives them an undulating flight pattern.

Q Flying seems like such a great thing. So why can't Ostriches and penguins and some other birds fly?

A Flying is a splendid ability, but it comes at a cost in terms of maintaining a huge keel on the sternum and large flight muscles. Most ratites, the group to which Ostriches belong, seem to have lost flight in order to increase their body size and the strength of their legs and feet. They can outrun most predators, and some can kick hard enough to disembowel dog-sized predators.

Penguins do have strong wings and strong pectoral muscles to power them. In a sense, they really do fly, except it's through water rather than air. Penguin bodies are streamlined as if for flight, which allows them to cut through the water with the least resistance.

But the requirements for rapid, powered swimming are different from those for flying. To dive deep and do the things penguins do in the water, and to survive the frigid air temperatures that many of them must endure, their bodies have huge fat supplies, heavy muscles, and extremely dense (and heavy!) feathers. That makes them too heavy to fly with such small wings.

Q If penguins, which all live in the Southern Hemisphere, are so good at flying underwater, why haven't
any northern birds, such as loons or puffins, lost their ability to fly?

A This may be because Antarctica and nearby islands provide low, safe places for penguins to nest, while Arctic
islands are steep-sided, so oceanic birds in the north had to
nest on more inaccessible cliffs.
Only one Northern Hemisphere
marine species lost the ability
to fly: the Great Auk. This bird
nested only on two or three low
islands and became extinct in
the midnineteenth century.

∨ ∨

Q Do birds sleep while in flight?

A There's evidence that some birds, especially long-
distance migrants and swifts, may occasionally sleep in
flight. Birds and marine mammals are the only warm-blooded
animals we know of that have "unihemispheric slow-wave
sleep" — that is, one side of their brain may sleep while the
other half is awake. Marine mammals apparently do this so
they can continue to swim and rise to the surface to breathe
as they sleep. Birds on the ground can sleep with one eye
open, instantly reacting to a predator's approach. This ability

may also allow them to sleep on the wing, which would be a handy skill for long-distance flights, although so far no one has proven this using brainwave monitoring, either under laboratory conditions or in nature.

∧∨

The Highs and Lows of Flight

Q How high can birds fly?

A The highest documented bird flight was of a group of Bar-headed Geese seen and heard crossing the Himalayas from India to central Asia at 29,000 feet (8,839 m). A Mallard was struck by an airplane over the Nevada desert at an altitude of 21,000 feet (6,400 m). Tiny Blackpoll Warblers sometimes migrate as high as 21,000 feet on their 2,300-mile (3,700 km) nonstop journey over the Atlantic Ocean. As far as we know,

HOW LOW CAN YOU GO?

🐦 **Observers at lighthouses** and other vantage points note that certain migrants commonly travel at altitudes of a few feet to a few hundred feet above sea or land. Sandpipers, Red-necked Phalaropes, pelicans, and various sea ducks have been seen flying so low they were visible only as they topped the waves.

this is the longest flight, both in distance and time, and the highest nonstop flight, of such small birds.

Radar observations indicate that birds on long-distance flights move along at higher altitudes than short-distance migrants. Advantageous tail winds of greater velocity are found higher up, along with cooler air that helps birds dissipate the heat they generate under the exertion of flying.

˅ ˅

Q What's the fastest bird? The slowest?

A The fastest bird may be the Peregrine Falcon, which has been clocked with various tools, including police radar, going at least 180 miles per hour (290 km per hour) and perhaps over 200 mph (322 kph). No one has clocked the White-throated Swifts that have successfully eluded peregrines — they're certainly slower than a diving peregrine but are great at evasive maneuvers.

If we count the slowest bird as one that can remain in flight for many minutes going zero miles per hour, I suppose we could give that distinction to various hummingbirds. If we count the slowest bird beating its wings in forward flight without stalling out, American Woodcocks and Eurasian Woodcocks have both been clocked at just 5 mph (8 kph) in courtship

flights. But American Woodcocks don't always fly so slowly; they've also been clocked at 13 mph (21 kph) and even 42 mph (68 kph).

v v v v v v v v v v v v v v v v v v v v v v v v v

Q How far can a bird fly without resting?

A In terms of normal migration, flight distance depends on the species. Many Ruby-throated Hummingbirds take off from the Texas and Louisiana gulf coasts and fly nonstop to the Yucatán Peninsula, a minimum of almost 600 miles (965 km) with nowhere to rest or feed en route. Blackpoll Warblers migrate over the Atlantic Ocean from the northeastern United States to Puerto Rico, the Lesser Antilles, or northern South America. This route averages 1,864 miles (2,999 km) over water, sometimes requiring a nonstop flight of up to 88 hours. To accomplish this flight, the Blackpoll Warbler nearly doubles its body mass before the trip and takes advantage of a shift in prevailing wind direction to direct it to its destination.

The longest nonstop flight ever recorded for a bird wearing satellite tags was a female Bar-tailed Godwit, a shorebird that flew 7,145 miles (11,496 km) from Alaska to New Zealand without taking a break for food or drink during the nine-day journey in October 2007.

NIGHT FLIGHTS

Radar studies have shown that nocturnal migrants, including most songbirds, fly at different altitudes at various times during the night. These birds generally take off shortly after sundown and rapidly reach maximum altitude. They remain high until about midnight, at which point they gradually descend until daylight. There is considerable variation, but most small birds appear to migrate at between 500 and 1,000 feet (152–305 m). Some nocturnal migrants (probably shorebirds) fly over the ocean at 15,000 or even 20,000 feet (4,572–6,096 m). Nocturnal migrants also fly slightly higher than diurnal migrants.

Flocks and Formations

Q Hawks are supposed to be solitary, but I saw a whole flock of them migrating. Do they band together for protection?

A Most hawks aren't gregarious. But during migration, many species, particularly the buteos (the group of hawks that are often seen "making lazy circles in the sky," as in the song "Oklahoma"), seek out rising columns of air called "thermal air currents" and "updrafts." When a hawk feels one of these air currents, it widens its wings and tail to provide maximum surface area and begins circling as the air column carries it up in a spiral.

The hawk climbs as high as the air will carry it, then pulls its wings back in something of an arrow shape, and glides forward in the direction it's migrating. Now it's steadily losing altitude, but it may cover quite a distance before reaching treetop height; and meanwhile, on a good day, it's found another thermal or updraft. Gliding between these rising currents can allow a hawk to cover hundreds of miles in a single day while using a minimum of energy.

A hawk may cover considerable ground seeking out thermals and updrafts, although experienced birds learn to search near shorelines and above pavement for a thermal, where the temperature is at least a few degrees warmer than nearby water or vegetated ground, causing the warmer air to rise, or to search near bluffs and tall buildings for an updraft, because such obstructions force oncoming winds upward. Air currents are invisible, but hawks can easily see exactly where a thermal or updraft is by watching for other hawks spiraling upward. They are drawn not to the other hawks but to the rising air. Over

time on a sunny morning, hundreds or thousands of hawks can be seen spiraling on the same thermals.

SEE ALSO: *pages 71 and 218 for more on hawk migration.*

Q Why do geese fly in a V?

A There are two advantages to a V formation. The first is aerodynamic. As a single goose flaps through the air, it disturbs the air by creating wingtip vortices. These vortices are generally undesirable because they create a downwash that increases drag on that wing. However, this downwash is also accompanied by an upwash that can help a bird flying behind and slightly above it, providing lift and reducing drag. This means that the second bird doesn't have to flap as hard or often to maintain the same speed as the first.

Although the birds trailing the first bird benefit most, the first bird also saves energy by having birds flying behind it, because their presence helps dissipate the upwash. In a long V, the birds saving the least energy are the one in front and the two pulling up the rear. That's why birds flying in formation shift positions fairly often. In one study, researchers monitored pelican heartbeats as the birds flew. They found that the heart rates of pelicans flying in formation were much lower than those of pelicans flying alone.

The second reason birds fly in a V is for the same reason that Air Force jets do — to more easily maintain visual contact with one another.

Q Every now and then when I'm going through farm country, I see a huge flock of songbirds wheeling through the air in what looks like a big black cloud. What are they, and why do they fly that way? How come they don't crash into each other?

A You're looking at a flock of European Starlings, famous the world over for that wondrous flying. When they get into that amazing flock formation, chances are there's a hawk nearby. Hawks are reluctant to get too close to one of these swarms, partly because of the unpredictable changes of direction the birds make and partly because the birds are so close together that the hawk risks colliding with one while grabbing another.

How do the birds within one of these swarms avoid each other? There are still mysteries involved, but we do know a few things thanks to high-speed photography, which allows us to slow the action. Birds seem to key in not on their immediate neighbors but on more distant birds, so the smoothness in changes happens the same way sports fans produce a "wave," by watching the people more in the distance and anticipating the right moment to act. No particular bird in a swarm is the leader. Any individual can begin a new maneuver, banking toward the center of the group, and that movement spreads through the flock like that wave.

Appendixes

Scientific Bird Names

Common Name	Taxonomic Name
Acorn Woodpecker	*Melanerpes formicivorus*
American Bittern	*Botaurus lentiginosus*
American Crow	*Corvus brachyrhynchos*
American Dipper	*Cinclus mexicanus*
American Goldfinch	*Carduelis tristis*
American Kestrel	*Falco sparverius*
American Robin	*Turdus migratorius*
American White Pelican	*Pelecanus erythrorhynchos*
American Woodcock	*Scolopax minor*
Anna's Hummingbird	*Calypte anna*
Apapane	*Himatione sanguinea*
Arctic Tern	*Sterna paradisaea*
Atlantic Puffin	*Fratercula arctica*
Australian Brush-turkey	*Alectura lathami*
Bald Eagle	*Haliaeetus leucocephalus*
Baltimore Oriole	*Icterus galbula*
Barn Swallow	*Hirundo rustica*
Bar-headed Goose	*Anser indicus*
Bar-tailed Godwit	*Limosa lapponica*
Barred Owl	*Strix varia*
Bewick's Wren	*Thryomanes bewickii*
Black-browed Albatross	*Thalassarche melanophris*
Blackburnian Warbler	*Dendroica fusca*
Black-capped Chickadee	*Poecile atricapillus*
Black-crowned Night-Heron	*Nycticorax nycticorax*
Black-footed Albatross	*Phoebastria nigripes*
Black-headed Duck	*Heteronetta atricapilla*

Common Name	Taxonomic Name
Black-headed Grosbeak	*Pheucticus melanocephalus*
Black-nest Swiftlet	*Aerodramus maximus*
Blackpoll Warbler	*Dendroica striata*
Black Swift	*Cypseloides niger*
Black Vulture	*Coragyps atratus*
Blue-footed Booby	*Sula nebouxii*
Blue Jay	*Cyanocitta cristata*
Bobolink	*Dolichonyx oryzivorus*
Boreal Chickadee	*Poecile hudsonica*
Brambling	*Fringilla montifringilla*
Broad-winged Hawk	*Buteo platypterus*
Brown-headed Cowbird	*Molothrus ater*
Brown Pelican	*Pelecanus occidentalis*
Brown Thrasher	*Toxostoma rufum*
Budgerigar	*Melopsittacus undulatus*
California Condor	*Gymnogyps californianus*
Calliope Hummingbird	*Stellula calliope*
Canada Goose	*Branta canadensis*
Cape May Warbler	*Dendroica tigrina*
Carolina Wren	*Thryothorus ludovicianus*
Cedar Waxwing	*Bombycilla cedrorum*
Chipping Sparrow	*Spizella passerina*
Cliff Swallow	*Petrochelidon pyrrhonota*
Club-winged Manakin	*Machaeropterus deliciosus*
Common Goldeneye	*Bucephala clangula*
Common Loon	*Gavia immer*
Common Nighthawk	*Chordeiles minor*
Common Raven	*Corvus corax*
Cooper's Hawk	*Accipiter cooperii*

Common Name	Taxonomic Name
Dark-eyed Junco	*Junco hyemalis*
Downy Woodpecker	*Picoides pubescens*
Eastern Bluebird	*Sialia sialis*
Eastern Kingbird	*Tyrannus tyrannus*
Eastern Meadowlark	*Sturnella magna*
Eastern Phoebe	*Sayornis phoebe*
Eastern Screech-Owl	*Megascops asio*
Edible-nest Swiftlet	*Aerodramus fuciphagus*
Emperor Penguin	*Aptenodytes forsteri*
Eurasian Collared-Dove	*Streptopelia decaocto*
Eurasian Kestrel	*Falco tinnunculus*
Eurasian Woodcock	*Scolopax rusticola*
European Pied Flycatcher	*Ficedula hypoleuca*
European Starling	*Sturnus vulgaris*
Evening Grosbeak	*Coccothraustes vespertinus*
Florida Scrub-Jay	*Aphelocoma coerulescens*
Fork-tailed Flycatcher	*Tyrannus savana*
Garden Warbler	*Sylvia borin*
Gilded Flicker	*Colaptes chrysoides*
Golden Eagle	*Aquila chrysaetos*
Gray Catbird	*Dumetella carolinensis*
Gray Jay	*Perisoreus canadensis*
Great Auk	*Pinguinus impennis*
Great Blue Heron	*Ardea herodias*
Great Crested Flycatcher	*Myiarchus crinitus*
Great Frigatebird	*Fregata minor*
Great Gray Owl	*Strix nebulosa*
Great Horned Owl	*Bubo virginianus*
Greater Prairie-Chicken	*Tympanuchus cupido*

Common Name	Taxonomic Name
Greater Roadrunner	*Geococcyx californianus*
Greater Sage-Grouse	*Centrocercus urophasianus*
Greater Yellow-headed Vulture	*Cathartes melambrotus*
Green Heron	*Butorides virescens*
Green Woodpecker	*Picus viridis*
House Finch	*Carpodacus mexicanus*
House Sparrow	*Passer domesticus*
House Wren	*Troglodytes aedon*
Iiwi	*Vestiaria coccinea*
Indigo Bunting	*Passerina cyanea*
Killdeer	*Charadrius vociferus*
Kirtland's Warbler	*Dendroica kirtlandii*
Lammergeier	*Gypaetus barbatus*
Laysan Albatross	*Phoebastria immutabilis*
Lesser Prairie-Chicken	*Tympanuchus pallidicinctus*
Lesser Yellow-headed Vulture	*Cathartes burrovianus*
Mallard	*Anas platyrhynchos*
Marsh Wren	*Cistothorus palustris*
Merlin	*Falco columbarius*
Mourning Dove	*Zenaida macroura*
New Caledonian Crow	*Corvus moneduloides*
Northern Cardinal	*Cardinalis cardinalis*
Northern Flicker	*Colaptes auratus*
Northern Mockingbird	*Mimus polyglottos*
Northern Saw-whet Owl	*Aegolius acadicus*
Northern Shrike	*Lanius excubitor*
Oilbird	*Steatornis caripensis*
Orange-collared Manakin	*Manacus aurantiacus*
Osprey	*Pandion haliaetus*

Common Name	Taxonomic Name
Ostrich	*Struthio camelus*
Pacific Golden-Plover	*Pluvialis fulva*
Peregrine Falcon	*Falco peregrinus*
Pileated Woodpecker	*Dryocopus pileatus*
Pine Warbler	*Dendroica pinus*
Piping Plover	*Charadrius melodus*
Prothonotary Warbler	*Protonotaria citrea*
Purple Martin	*Progne subis*
Pygmy Nuthatch	*Sitta pygmaea*
Red-capped Manakin	*Pipra mentalis*
Red-headed Woodpecker	*Melanerpes erythrocephalus*
Red Knot	*Calidris canutus*
Red-tailed Hawk	*Buteo jamaicensis*
Red-tailed Tropicbird	*Phaethon rubricauda*
Red-winged Blackbird	*Agelaius phoeniceus*
Ring-billed Gull	*Larus delawarensis*
Rose-breasted Grosbeak	*Pheucticus ludovicianus*
Ruby-throated Hummingbird	*Archilochus colubris*
Ruffed Grouse	*Bonasa umbellus*
Rufous Hummingbird	*Selasphorus rufus*
Sandhill Crane	*Grus canadensis*
Scarlet Tanager	*Piranga olivacea*
Sharp-shinned Hawk	*Accipiter striatus*
Sharp-tailed Grouse	*Tympanuchus phasianellus*
Snail Kite	*Rostrhamus sociabilis*
Snow Geese	*Chen caerulescens*
Snowy Owl	*Bubo scandiacus*
Song Sparrow	*Melospiza melodia*
Sooty Grouse	*Dendragapus fuliginosus*

Common Name	Taxonomic Name
Sooty Shearwater	*Puffinus griseus*
Sooty Tern	*Onychoprion fuscatus*
Steller's Jay	*Cyanocitta stelleri*
Swainson's Hawk	*Buteo swainsoni*
Swamp Sparrow	*Melospiza georgiana*
Tree Swallow	*Tachycineta bicolor*
Tufted Titmouse	*Baeolophus bicolor*
Turkey Vulture	*Cathartes aura*
Varied Thrush	*Ixoreus naevius*
Wandering Albatross	*Diomedea exulans*
Western Grebe	*Aechmophorus occidentalis*
Western Scrub-Jay	*Aphelocoma californica*
Whip-poor-will	*Caprimulgus vociferus*
White-collared Manakin	*Manacus candei*
White-crowned Sparrow	*Zonotrichia leucophrys*
White Ibis	*Eudocimus albus*
White Tern	*Gygis alba*
White-throated Sparrow	*Zonotrichia albicollis*
Whooping Crane	*Grus americana*
Wilson's Snipe	*Gallinago delicata*
Wilson's Storm-Petrel	*Oceanites oceanicus*
Winter Wren	*Troglodytes troglodytes*
Wood Duck	*Aix sponsa*
Wood Stork	*Mycteria americana*
Wood Thrush	*Hylocichla mustelina*
Worm-eating Warbler	*Helmitheros vermivorum*
Yellow-bellied Sapsucker	*Sphyrapicus varius*
Yellow-rumped Warbler	*Dendroica coronata*
Yellow Warbler	*Dendroica petechia*

Glossary

anting behavior. The rubbing of ant bodies against the feathers or sitting on an anthill allowing ants to crawl through the feathers. Ant bodies are covered with a bitter chemical called formic acid, which may afford some protection from mites and lice. People have reported birds anting with items such as mothballs, cigarette butts, and onions.

aspect ratio. How long a wing is compared to how wide it is. This determines how a bird soars. High aspect ratios (long, narrow wings) allow flight at high speeds at low altitudes, while being extremely maneuverable. Low aspect ratios (broader wings) are adapted to slow speed, high altitude, effortless flight. Bird wings with aspect ratios at either extreme make taking off from the ground without a stiff headwind difficult.

austral migrants. Birds that breed in the Southern Hemisphere and migrate north for their winter.

brood parasites. Birds that do not build a nest or care for their young directly, instead searching out host nests in which to lay their eggs.

brood patch. A bare spot on the belly or chest where a bird's body heat warms the eggs.

cloaca. The opening chamber to the intestines, ureters, and sex organs.

contour feathers. The outer feathers that keep moisture and wind out and streamline a bird.

countersinging. When neighboring birds sing in response to each other.

distraction display. Feigning an injury with loud calls and drooping wings to draw predators away from the nest.

diurnal migrants. Birds that migrate during the day.

down feathers. The inner feathers that trap air, providing insulation to hold body heat inside.

generalists. Bird species that can obtain everything they need from a variety of habitats.

gizzard. A muscular chamber of the stomach that mashes the food.

glottis. A fairly large opening on the bottom of the bird's mouth where the trachea begins.

guano. The fecal and urinary waste of birds, especially seabirds, which contains a lot of uric acid and is collected for nitrogen- and phosphorus-rich fertilizers.

innate behavior (also instinctive behavior). A behavior that birds do in a particular situation without learning or trying that behavior beforehand.

indeterminate layer. A bird that will continue laying more eggs for a long time if its eggs are removed one by one.

insectivores. Birds that feed primarily on insects.

isotherm. The "line" visible on a weather map where temperatures average a particular temperature.

kettle. A swirling mass of hawks numbering anywhere from several to many thousands.

lacrimal gland. A tear gland at the base of the nictitating membrane that maximizes lubrication during blinking.

lek. An area where males gather to display and attract mates.

migrational orientation. Taking a direction and keeping it during a leg of migration.

mimids. Birds in the mockingbird family, including thrashers and catbirds.

neotropical migrants. Birds that spend winter in the Caribbean or Central or South America and breed in temperate North America.

nictitating membrane. A semitransparent inner eyelid that helps protect bird eyes and keep them moist.

nocturnal migrants. Birds that migrate at night.

owl pellets. The bones, fur, teeth, and other indigestible matter that an owl regurgitates in a dense, compact mass.

perching reflex. A mechanism whereby the stretched tendons of the lower leg automatically flex the foot around the branch and lock it in place while a bird is perching.

proventriculus. A glandular chamber of the stomach where powerful acids start dissolving the food.

rictal bristles. Specialized feathers found near the beaks of some insect-eating birds that provide tactile sensitivity and may sometimes help funnel insects into the bird's mouth.

shadowboxing. A nontechnical term for when a bird fights with its own reflection in a window or mirror.

socially monogamous. Birds that defend a territory and raise young as a pair, but may sometimes mate with other individuals.

specialists. Bird species that have restricted habitat, nesting, or food needs.

syrinx (also song box). The structure used by birds to produce sound, located where the trachea branches into the bronchial tubes. The separate branches and sets of muscles allow many species to produce harmony with their own voice.

tubenose. A name for some ocean birds, such as albatrosses, that have huge glands that excrete salt at the end of "tubes" on their upper bills.

unihemispheric slow-wave sleep. A state in which one side of the brain may be asleep while the other half is awake.

uropygial gland (also preen gland). A gland found at the base of the tail that produces drops of oil used by birds to keep their feathers in good condition.

vanes. The webbed sides of the feather.

Wulst. The area of the brain that processes information from both eyes to provide stereoscopic vision.

Zugunruhe. Migratory restlessness in response to changing day length and/or the changing angle of the sun in the sky. Virtually all migratory species experience this urge, including individuals that have been hatched and reared in captivity.

Resources

All About Birds
An online resource produced by the Cornell Lab of Ornithology; a wealth of information about each species of North American birds, including photos and sound recordings
www.allaboutbirds.org

Birds of North America Online
Produced by the Cornell Lab of Ornithology and the American Ornithologists' Union, is an in-depth survey of each species of bird that breeds in North America, with an extensive bibliography for each.
www.bna.birds.cornell.edu

Field Guides

Dunn, Jon L., and Jonathan Alderfer. *National Geographic Field Guide to the Birds of North America,* Fifth edition. Washington, D.C.: National Geographic Society, 2006.

Kaufman, Kenn. *Kaufman Field Guide to Birds of North America.* Boston: Houghton Mifflin, 2000.

Peterson, Roger Tory. *A Field Guide to the Birds of Eastern and Central America.* Boston: Houghton Mifflin, 2002.

Peterson, Roger Tory. *A Field Guide to Western Birds.* Boston: Houghton Mifflin, 1998.

Sibley, David Allen. *The Sibley Guide to Birds.* New York: Knopf, 2000.

Learning and Understanding Bird Song

These books, each with an accompanying CD, contain a wealth of information about how and why birds sing.

Elliott, Lang. *Music of the Birds: A Celebration of Bird Song.* Boston: Houghton Mifflin, 1999.

Kroodsma, Don. *The Singing Life of Birds.* Boston: Houghton Mifflin, 2005.

Colver, Kevin, and L. Elliott. *Know Your Bird Sounds: Common Western Species.* Mechanicsburg, Penn.: Stackpole, 2008.

Elliott, Lang. *Know Your Bird Sounds, Volume 1: Yard, Garden, and City Birds.* Mechanicsburg, Penn.: Stackpole, 2004.

Elliott, Lang. *Know Your Bird Sounds, Volume 2: Birds of the Countryside.* Mechanicsburg, Penn.: Stackpole, 2004.

The following CD sets are primers explaining what elements to key in on when learning to identify bird songs and calls.

Walton, Richard K., and R. Lawson. *Birding by Ear: Eastern and Central North America.* Boston: Houghton Mifflin, 2002.

Walton, Richard K., and R. Lawson. *Birding by Ear: Western North America.* Boston: Houghton Mifflin, 1999.

Walton, Richard K., and R. Lawson. *More Birding by Ear: Eastern and Central North America.* Boston: Houghton Mifflin, 2000.

Recommended Reading

Chu, Miyoko. *Songbird Journeys: Four Seasons in the Lives of Migratory Birds.* New York: Walker & Company, 2007.

Cornell Lab of Ornithology. *Handbook of Bird Biology.* Princeton, N.J.: Princeton University Press, 2004.

Erickson, Laura. *101 Ways to Help Birds.* Mechanicsburg, Penn.: Stackpole, 2006.

Kaufman, Kenn. *Lives of North American Birds.* Boston: Houghton Mifflin, 2001.

Kaufman, Kenn. *Kingbird Highway: The Biggest Year in the Life of an Extreme Birder.* Boston: Houghton Mifflin, 2006.

Kress, Stephen W. *The Audubon Society Guide to Attracting Birds: Creating Natural Habitats for Properties Large and Small,* Second edition. Ithaca, N.Y.: Cornell University Press, 2006.

Weidensaul, Scott. *Living on the Wind: Across the Hemisphere with Migratory Birds.* New York, NY: North Point Press, 2000.

Sharing Your Bird Sightings

eBird

A splendid online resource that compiles birding lists from North America and the world over, giving our individual sightings a huge level of importance beyond our personal fun, providing helpful data for understanding birds, migration, and conservation issues.
www.ebird.edu

Project FeederWatch

A winter-long survey of birds that visit feeders all over North America, helping scientists track broad-scale movements of winter bird populations and long-term trends in bird distribution and abundance.
www.birds.cornell.edu/pfw

NestWatch

Aims to provide a unified nest-monitoring scheme to track reproductive success for all North American breeding birds: citizen scientists submit their nest records to NestWatch's online database where their observations are compiled with those of other participants in a continent-wide effort to better understand and manage the impacts of environmental change on bird populations.
www.nestwatch.org

The Christmas Bird Count

Longest-running wildlife census to assess the health of bird populations, conducted from mid-December through early January and sponsored by Audubon; data provides a look at bird populations in early winter throughout the Americas.
www.audubon.org/bird/cbc

The Great Backyard Bird Count

This annual four-day event engages bird-watchers in counting birds to create a real-time snapshot of where the birds are across the continent in mid-winter, often at the peak of northern bird invasions into the more populated areas of North America; sponsored by the Cornell Lab of Ornithology and Audubon.
www.birdsource.org/gbbc

Index

Page numbers in *italics* indicate illustrations.

R

radar tracking, 226–27, 359
  Next Generation Radar
    (NEXRAD), 227
rails, 229
raptors, 72, 122, 139, 266
ravens, 115, 168–69, 202, 242
  Common, 168
recording bird sounds, 58
Red Knot, 132
redpolls, 11, 13, 176, 306
respiratory system, 281–83
  breathing rates, 284
  getting oxygen at high altitudes,
    283
  lung capacity and function,
    281–82
ricebirds, 17
robins, American, 25, 89, 126, 132,
    135, 264
  albino, 338–39
  attaching window reflections,
    87–88
  danger from pesticides, 122–25
  diet sources, 121
  eyesight, 315
  gender differences, 241
  instinctive behavior, 170–73
  migration, 208, 210, 218, 232
  nesting behavior, 240–41, 248,
    250, 262, 304
  vocalization, 184–85, 189,
    202–3, 241
roosting, 250–51

S

sage-grouse, 117
  Greater, 132, 134, 238

salmonella, 15, 17, 42, 110
saltwater, drinking, 286, 288
sandpipers, 324, 356
sapsuckers, Yellow-bellied, 10, 200,
    248–49
seagulls. *See* gulls
sense of smell, 319–22
sexual dimorphism, 242
shearwaters, 288, 319
  Sooty, 209
shorebirds, 61, 156, 265, 337
shrikes, 194
  Loggerhead, 244
  Northern, 296
sick birds, 41–42, 108, 110–12, 114
sighting records, 63
siskins, 11–13, 176
skeletons, lightweight, 353
Skinner, B. F., 36
snipes, 324
  Wilson's, 205
social-weavers, African, 251
sonar or echolocation, 318–19
songbirds, 273–74, 317, 320, 324,
    351
  mating rituals, 34, 236
  migration, 72–73, 131, 214, 220,
    225
  territorial needs, 243
  thick flock formation, 362
sparrows, 11, 289
  Barn, 248, 255–56
  Chipping, 185–186, 189, 202,
    251
  House, 34, 113, 134–35, 258–61,
    339
  Savannah, 184
  Song, 126, 188, 194–95, 243